Martin H. Evans

Received for Review
3rd January 1975

Bioactive Compounds from the Sea

Marine Science

Series Editor

Donald W. Hood
Marine Science Institute
University of Alaska
Fairbanks, Alaska

Volume 1 **Bioactive Compounds from the Sea**
Edited by Harold J. Humm
and Charles E. Lane

in preparation

Internal Gravity Waves in the Ocean
Jo Roberts

Bioactive Compounds from the Sea

Edited by

Harold J. Humm

Department of Marine Science
University of South Florida
St. Petersburg, Florida

and

Charles E. Lane

Rosenstiel School of Marine and Atmospheric Science
University of Miami
Miami, Florida

MARCEL DEKKER, INC. New York 1974

MARCEL DEKKER, INC.

305 East 45th Street, New York, New York 10017

LIBRARY OF CONGRESS CATALOG CARD NUMBER: 73-90764

ISBN: 0-8247-6090-5

Current printing (last digit):
10 9 8 7 6 5 4 3 2 1

PRINTED IN THE UNITED STATES OF AMERICA

CONTENTS

List of Contributors vii
Foreword xi
Preface xiii

Chapter 1 SOME ASPECTS OF THE BIOLOGICAL ACTIVITY OF CRUDE NEMATOCYST TOXIN FROM CHIRONEX FLECKERI 1
Robert Endean and Lyn Henderson

I. Introduction 1
II. Materials and Methods 3
III. Results 5
IV. Discussion 9
References 12

Chapter 2 THE BIOLOGICAL ORIGIN AND TRANSMISSION OF CIGUATOXIN 15
Albert H. Banner

I. Biological Patterns of Ciguatoxic Fishes 17
II. Origin and Transmission of Ciguatoxin 24
III. The Ecology of Outbreaks 28
IV. Conclusions 33
References 34

Chapter 3 MODES OF ACTION AND IDENTITIES OF PROTEIN CONSTITUENTS IN SEA URCHIN TOXIN 37
George A. Feigen, Lahlou Hadji, and John E. Cushing

I. Introduction 37
II. Preparation 38
III. Pharmacological Experiments on Isolated Mammalian Tissues 44
IV. Biochemical Studies on the Enzymatic Formation of Plasma Kinins 52
V. Discussion 89
VI. Range of Action of SUT Components 89
VII. Identification of SUT Components by Immunoelectrophoresis 92
VIII. Conclusions 95
References 96

Chapter 4 ULTRASTRUCTURE OF *PHYSALIA* NEMATOCYSTS 99
William H. Hulet, J.L. Belleme, G. Musil, and Charles E. Lane

I. Material 100
II. Methods 100
III. Results and Discussion 101
References 113

Chapter 5 DIFFERENCES BETWEEN THE ACTIONS OF TETRODOTOXIN AND SAXITOXIN 115
C.Y. Kao

I. Introduction 115
II. Cellular Actions 116
III. Systemic Actions 119
References 121

Chapter 6 NEMATOCYST TOXINS OF COELENTERATES 123
Charles E. Lane

Text 123
References 134

Chapter 7 SOURCE OF THE TOXICITY OF PUFFER FISHES, GENUS *SPHOEROIDES* 139
Edward Larson, Jack Grossman, John Humphries, and John Klinovsky

I. History 139
II. Methods 140
III. Results 141
IV. Discussion 141
V. Summary 148
References 148

Chapter 8 SOME PHYSIOLOGICAL PROPERTIES OF DINOFLAGELLATE TOXINS 151
Dean F. Martin and George M. Padilla

I. Introduction 151
II. *Prymnesium parvum* Toxins 153
III. Methods 160
IV. Results and Discussion 166
V. Concluding Remarks 171
References 171

Chapter 9 PARTIAL PURIFICATION AND BIOLOGICAL PROPERTIES OF AN EXTRACT OF THE GREEN SPONGE, HALICLONA VIRIDIS 175
Robert E. Middlebrook, Carl H. Snyder, Arsenio Rodriguez Mercado, and Charles E. Lane

I. Introduction 175
II. Materials and Methods 176
III. Results 177
IV. Discussion 181
References 182

Chapter 10 PHYSIOLOGICALLY ACTIVE SUBSTANCES FROM ECHINODERMS 183
George D. Ruggieri and Ross F. Nigrelli

I. Effects on Fundulus heteroclitus 187
II. Effects on Sea Urchin Gametes and Development 188
III. Effects on Tissue Culture Cells 190
IV. Antifungal Effects 191
V. Phylogenetic Relationships of Echinoderms 192
References 194

Chapter 11 BIOLOGICAL ACTIVITY EXHIBITED BY SEAWEED EXTRACTS 197
T.L. Senn

I. Biological Activity 200
II. Methods of Procedure: Biological Activity 200
Addendum 201
References 205

Chapter 12 DISTRIBUTION OF SEA SNAKES IN SOUTHEAST ASIA AND THE FAR EAST AND CHEMISTRY OF VENOMS OF THREE SPECIES 207
Anthony T. Tu

I. Sea Snake Collecting 207
II. Biology 211
III. Toxicology 214
IV. Chemistry 215
V. Immunology 221
VI. Discussion 221
References 229

Chapter 13 MARINE PROSTAGLANDINS 231
Alfred J. Weinheimer

Text 231
References 238

Author Index 239
Subject Index 247

LIST OF CONTRIBUTORS

Numbers in parentheses indicate the pages on which the authors' contributions begin.

ALBERT H. BANNER (15), Department of Zoology, Hawaii Institute of Marine Biology, University of Hawaii, Honolulu, Hawaii

J.L. BELLEME (99), Electron Microscopy Laboratory, Veterans Administration Hospital, Miami, Florida

JOHN E. CUSHING (37), Department of Biological Sciences, University of California at Santa Barbara, Santa Barbara, California

ROBERT ENDEAN (1), Department of Zoology, University of Queensland, Brisbane, Queensland, Australia

GEORGE A. FEIGEN (37), Department of Physiology, School of Medicine, Stanford University, Stanford, California

JACK GROSSMAN (139), Department of Biology, University of Miami, Coral Gables, Florida and The Miami Seaquarium, Miami, Florida

LAHLOU HADJI (37), Department of Physiology, School of Medicine, Stanford University, Stanford, California

LYN HENDERSON (1),* Department of Zoology, University of Queensland, Brisbane, Queensland, Australia

WILLIAM H. HULET (99),† Division of Undersea Medicine, School of Medicine, University of Miami, Miami, Florida

*Present address: Department of Paramedical Studies, Queensland Institute of Technology, Brisbane, Queensland, Australia

†Present address: Marine Medicine Division, Marine Biomedical Institute, University of Texas, Galveston, Texas

JOHN HUMPHRIES (139), Department of Biology, University of Miami, Coral Gables, Florida and The Miami Seaquarium, Miami, Florida

C.Y. KAO (115), Department of Pharmacology, State University of New York, Downstate Medical Center, Brooklyn, New York

JOHN KLINOVSKY (139), Department of Biology, University of Miami, Coral Gables, Florida and The Miami Seaquarium, Miami, Florida

CHARLES E. LANE (99, 123, 175), Rosenstiel School of Marine and Atmospheric Science, University of Miami, Miami, Florida

EDWARD LARSON (139), Department of Biology, University of Miami, Coral Gables, Florida and The Miami Seaquarium, Miami, Florida

DEAN F. MARTIN (151), Department of Chemistry, University of South Florida, Tampa, Florida

ARSENIO RODRIGUEZ MERCADO (175), Rosenstiel School of Marine and Atmospheric Sciences, University of Miami, Miami, Florida

ROBERT E. MIDDLEBROOK (175),* Department of Chemistry, University of Miami, Coral Gables, Florida

G. MUSIL (99), Electron Microscopy Laboratory, Veterans Administration Hospital, Miami, Florida

ROSS F. NIGRELLI (183), Osborn Laboratories of Marine Sciences, New York Aquarium, New York Zoological Society, Brooklyn, New York

GEORGE M. PADILLA (151), Department of Physiology and Pharmacology, Duke University Medical Center, Durham, North Carolina

GEORGE D. RUGGIERI (183), Osborn Laboratories of Marine Sciences, New York Aquarium, New York Zoological Society, Brooklyn, New York

*Present address: Department of Marine Sciences, University of Puerto Rico, Mayaguez, Puerto Rico

T.L. SENN (197), Department of Horticulture, Clemson University, Clemson, South Carolina

CARL H. SNYDER (175), Department of Chemistry, University of Miami, Coral Gables, Florida

ANTHONY T. TU (207), Department of Biochemistry, Colorado State University, Fort Collins, Colorado

ALFRED J. WEINHEIMER (231), Department of Chemistry, University of Oklahoma, Norman, Oklahoma

FOREWORD

Tales of exploration, the drama and the legend of the high seas have captured the fancy of persons of all ages since the beginning of time. The compelling romance of the waves and shore has plied the common soul of mankind from time immemorial; yet, only in the sliver of the past few decades has the literature begun to address the equally boundless domain of the science of the sea.

The ocean in many respects is a gargantuan pacemaker which vitally regulates the function and health of the world environment. Stripped of its bounty from the sea, the earth alone could not constitute a viable biological habitat. Conversely, an ocean that is limited in its function by any adverse mechanisms decreases the earth's capacity to sustain its terrestrial community.

Not only is the ocean itself of great importance to man, but its coastal margin is under extreme pressure for the location of cities and technological amenities and for the extraction of raw materials commercially valuable to a growing society. Man is unquestionably threatening the quality of the coastal zone in many countries of the world. The mirror tide of his advancing demands for ocean uses is relentlessly pounding the beaches of the civilized world.

Despite Man's heavy exploitation of the ocean and his integral dependence upon it for continued existence on this planet, our understanding of ocean and coastal processes is paradoxically and appallingly limited. We have very little knowledge of how much stress the ocean environment can withstand before significant deterioration or destruction occurs. To ensure and improve our life as we know it on earth, the limitations of the ocean must be fundamentally respected and its resources utilized not in a sense of unthinking conquest, but subordinate to the survival requirements of its own natural system.

Our ultimate scientific aim of engendering intelligent decisions on ocean resource utilization must then be to fully understand and plan compatibly within the framework of the complex interaction of the ocean and that critical region where land meets sea. It is to this purpose that the Marine Science series is dedicated.

Scientific knowledge of the ocean and its shorelines must be acquired, intensively analyzed, and disseminated to all the public -- laymen and scientists alike. Our Marine Science series is undertaken in this direction. We hope to emphasize the interdisciplinary science of oceanography relevant to man; however, single-discipline material will be included in our publications to lend comprehensiveness of subject matter. We hope our readership will find benefit in these informative objectives. Your criticism and comments will always be welcome.

Donald W. Hood

PREFACE

A grant from the North Miami Hospital Foundation enabled the Marine Science Institute (now Department of Marine Science) of the University of South Florida to hold a symposium, Physiologically Active Compounds from Marine Organisms, on November 18 and 19, 1971, in St. Petersburg.

Fourteen scientists of international reputation accepted invitations to present a paper on their latest research at this symposium. Twelve of these papers are presented in this volume.

The editors and participants in the symposium are grateful to members of the North Miami Hospital Foundation and in particular to Mrs. Ralph Waldo Miner, Dr. Richard Gubner, and Dr. Alfred H. Lawton.

Much credit for the success and smooth running of the symposium is due Patricia Archer and Donna Mengerink for their faithful and effective management of the detailed arrangements so essential to the success of this endeavor.

Harold J. Humm

Charles E. Lane

Bioactive Compounds from the Sea

Chapter 1

SOME ASPECTS OF THE BIOLOGICAL ACTIVITY OF CRUDE NEMATOCYST TOXIN FROM CHIRONEX FLECKERI

Robert Endean and Lyn Henderson*

Department of Zoology
University of Queensland
Brisbane, Queensland, Australia

I. INTRODUCTION

In tropical Australian waters, numerous fatalities and near fatalities have resulted from contact with the nematocyst-laden tentacles of the jellyfish *Chironex fleckeri* (Cleland and Southcott, 1965; Barnes, 1966). The biological activity of extracts of material released from the ruptured nematocysts of this cubomedusan (Fig. 1) has been studied by Endean et al. (1969) and by Endean and Henderson (1969). It has been shown that this material elicits sustained contractures of skeletal, respiratory, and extravascular smooth musculature of rats and that it has a marked action on the cardiovascular system of rats. The hearts of rats exposed in situ and injected with concentrated doses of extracts showed bradycardia associated with a progressive decrease in the extent of ventricular relaxation with each successive cardiac cycle culminating in the paralysis of each heart in systole. Less concentrated doses of extracts did not paralyze heart musculature but elicited a marked hypertension associated with pulmonary edema. In view of these results, it was decided to ascertain whether the extracts affected vascular smooth muscle and capillary permeability as well as cardiac activity. Also, it was decided to determine whether isolated and perfused hearts of rats responded to the extracts in a similar way to exposed hearts which were under hormonal and nervous regulation, and whether extracts had any effect on action potentials in peripheral nerves.

*Present address: Department of Paramedical Studies, Queensland Institute of Technology, Brisbane, Queensland, Australia

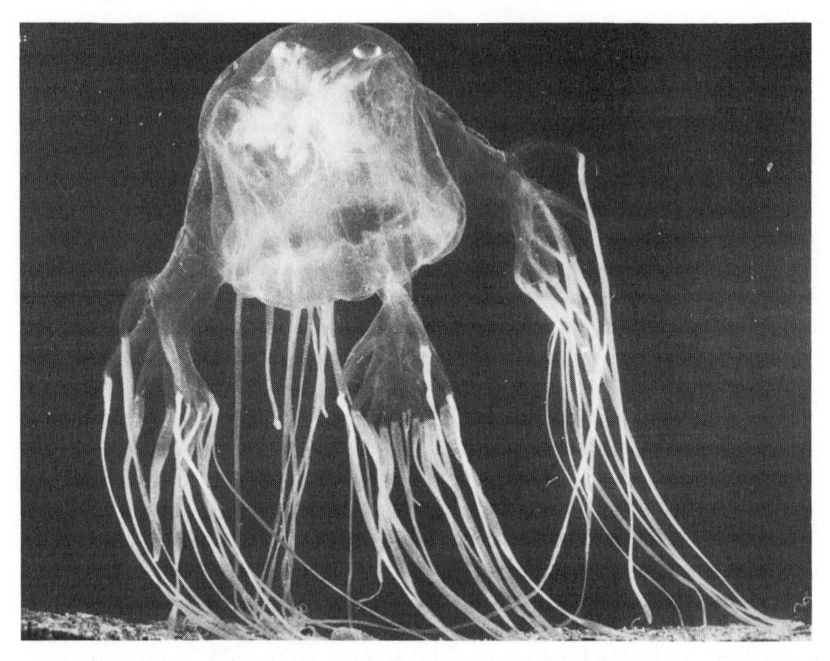

FIG. 1. Photograph of living specimen of Chironex fleckeri. The bell diameter is approximately 16 cm.

Endean and Henderson (1969) have shown that the nematocyst toxin extracts affected calcium ion movements across muscle fiber membranes. Exposure to adequate concentrations of the extracts caused an initial influx of calcium ions into barnacle musculature and this influx was associated with a sustained contracture of the musculature. Subsequently, there was a marked efflux of calcium ions, which appeared to be responsible for decline of the contracture and paralysis. In view of this, it was decided to investigate whether the toxin extracts affected the ability of the sarcoplasmic reticulum of mammalian muscles to accumulate calcium ions. Also, it was decided to ascertain whether the action of extracts on vertebrate muscle could be blocked by prior disruption of the transverse tubular system of the muscles or by prior exposure of the muscles to the action of tetrodotoxin or procaine.

It was shown earlier (Endean et al., 1969) that the crude nematocyst extracts possessed dermonecrotic activity, and it is known that full thickness skin necrosis can result from contact with Chironex nematocysts (Kingston and Southcott, 1960).

However, the extracts possess no proteolytic activity or phospholipase A activity, and the reason for the observed cytolytic activity is not known. In an attempt to throw light on this aspect, the L-amino acid oxidase activity of the extracts was examined.

II. MATERIALS AND METHODS

Saline extracts (0.9%) of toxic material from the nematocysts of C. fleckeri were obtained as described in a previous paper (Endean et al., 1969).

Lagendorff's method as modified by Gunn (1913) for the preparation of an isolated perfused heart was followed. The hearts of rabbits, guinea pigs, and rats were used. A thread was attached to the apex of each heart and activity of the cardiac musculature was monitored by connecting the thread via a pully system to a transducer (type A) linked to a Physiograph (E. and M. Instrument Co.). Each heart was exposed to measured amounts of nematocyst toxin extract.

Changes of pressure within an isolated perfused artery were measured according to the method of Waugh (1962). Portions of the mesenteric arteries from dogs and rabbits were used. The mesentery containing the arterial trunk was excised from each freshly killed animal and segments of artery 2-3 cm in length and 0.3-0.5 mm in external diameter were taken from it. Pressure change was measured with a Bourdon-type pressure transducer (E. and M. Instrument Co.), which was calibrated to read from 0-300 mm Hg pressure in 20 mm intervals. Constant flow perfusion at a rate of 0.8 ml per min was obtained by employing a Sage syringe pump, Model 255-1. Perfusion solutions used included mammalian Ringer's solution and 144 mM KCl solution prepared according to Svihovec and Raskova (1967). Test solutions used included adrenaline and nematocyst toxin extracts.

A partially desheathed nerve preparation was made by killing a toad (Bufo marinus) and exposing the sciatic nerve from the spinal cord to the knee. The nerve was then dissected out and transferred to a wax dish containing toad Ringer's solution. Threads were tied to the upper and lower parts of the nerve and the nerve was pinned out by these threads. A thread was also tied to the peroneal nerve, and this was held while an incision was made at the junction of the sciatic nerve and the peroneal nerve and the peroneal nerve stripped away from the sciatic nerve for a distance of approximately 1.5 cm. The partially desheathed nerve was placed in a perspex chamber containing electrodes for stimulating the nerve and for recording the action potentials produced. The preparation was arranged so that drops of Ringer's solution or nematocyst toxin extracts could be applied to the desheathed portion of the trunk. A tightly fitting lid was placed over the chamber and the preparation was kept moist with toad Ringer's solution. The nerve was stimulated by a Palmer square wave stimulator. Shocks of 2.5 msec duration at intervals of 8 sec, and at a nominal voltage of 2 V, were used. Action potentials were recorded by a Tektronix Type 502 dual-beam oscilloscope.

For capillary leakage studies, male rats were anaesthetized with ether and injected intravenously via the caudal vein with Evan's Blue, 2% w/v in 0.9% saline. An area approximately 3 x 4 cm was shaved on each rat's back and subcutaneous injections of the following solutions in 0.9% saline were administered here as required: 0.2 ml of 800 mM histamine; 0.5 ml of 200 mM EDTA; and 0.1 ml of 0.9% saline extracts containing the toxic material released from approximately 1×10^5 nematocysts.

Similar amounts of venom were also injected intraperitoneally and intravenously. The resulting lesions (if any) in the shaved areas were photographed.

Attempts to block the action of toxin extracts on rat phrenic nerve diaphragm preparations and rat skeletal muscle preparations by prior application of procaine (DHA) and on rat diaphragm preparations by prior application of tetrodotoxin (Sankyo) were made. The procaine concentration used on each rat phrenic nerve diaphragm preparation immersed in 50 ml of Kreb Ringer's solution was 0.5 μg/ml and on a rat pectoralis minor preparation was 0.75 μg/ml. Tetrodotoxin was used on rat phrenic nerve diaphragm preparations at a concentration of 4 μg/ml. Toxin extracts of 0.1 ml, each containing the contents of approximately 1.4×10^6 nematocysts, were used to elicit activity in the diaphragm preparations, and 0.1 ml extracts, each containing the contents of approximately 2.8×10^6 nematocysts, were used to elicit activity in the pectoralis minor preparations.

Preparations of the sartorius muscles of the frog *Hyla caerulea* were bathed in normal amphibian Ringer's solution in an organ bath and connected to a lever in order to record muscle movements. The preparations were stimulated at intervals of 5 sec with shocks of 1 msec duration and at a nominal voltage of 20 V until a steady response was given. The fluid was then drained from the bath and replaced with a Ringer's solution containing 750 mM glycerol. After standing for 1 hr exposed to the glycerol Ringer's solution, each preparation was stimulated electrically as before until a steady response was obtained. Again the bath was drained and the glycerol Ringer's solution was then replaced with normal amphibian Ringer's solution. When the musculature became refractory to stimulation, nematocyst toxin extracts were added to the bath. Different preparations were exposed to the toxin for periods ranging from several minutes to 1 hr. In some cases the amphibian Ringer's solution was then replaced with 10 mM caffeine.

Fragmented sarcoplasmic reticulum (FSR) freed of particulate material from mitochondria was prepared by the method of Ikemoto et al. (1968) using muscle obtained from the back and hind limbs of a mouse. To increase the longevity of the preparation, it was suspended in a medium containing 0.1 M KCl, 10 mM histidine (pH 7.0), and 0.3 M sucrose (Fuchs, 1969). The protein concentration of the FSR was estimated by reading the optical density at 280 nm with a Beckman DB spectrophotometer and comparing it with the optical density of a standard bovine serum albumen curve. Calcium uptake by the FSR from a medium containing 45 Ca prepared according to the method of Fuchs (1969) was studied. Seven samples, each of 0.05 ml and containing 1 mg FSR protein, were incubated at 37°C with 2 ml of this medium for periods of 1, 2, 4, 6, 8, 10 and 20 min, respectively. The samples were then filtered using a Millipore filter Type HA (0.45 μm pore size). The Millipore filters were dried overnight at 37°C then placed in a toluene scintillator (4 parts scintillator to 5 parts Triton X 100) and counted in an EKCO liquid scintillation counter. To test the effect of *Chironex* toxin extracts on calcium uptake 0.01 ml of toxin extract containing the contents of 1×10^5 nematocysts was incubated with 0.7 ml of FSR at 37°C for 10 min. Seven 0.05 ml samples of the treated FSR were incubated with 45 Ca medium for periods of 1, 2, 4, 6, 8, 10 and 20 min, respectively, and treated in the same way as the control samples.

L-Amino acid oxidase activity was investigated using the chromatographic method of Dimitrov and Kankonkar (1968).

III. RESULTS

The addition to the fluid perfusing the isolated heart of a guinea pig of 0.5 ml of toxin extract containing material released from 1×10^5 nematocysts resulted in the following sequence of events (Fig. 2): modifications in the amplitudes of

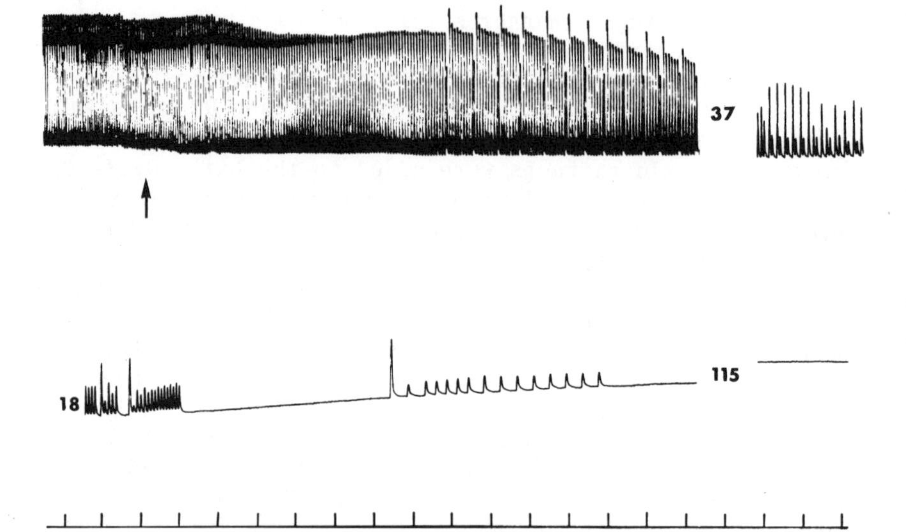

FIG. 2. Action of nematocyst toxin extract on isolated heart of guinea pig. Arrow indicates addition of 0.5 ml of toxin extract containing the contents of approximately 100,000 nematocysts to the perfusion medium. Intervals of 5 sec are indicated on time base. The duration in seconds of each gap in record is indicated.

cardiac contractions, bradycardia, a progressive decrease in the extent of ventricular relaxation with each successive cardiac cycle, a sustained paralysis of the ventricular musculature in systole, and paralysis of the atrial musculature. Typically, the peak of the sustained contracture of the ventricular musculature was reached approximately 5.5 min after the toxin was added to the perfusion medium. A slow relaxation of the ventricular musculature began after the lapse of another 3 to 4 min. However, recovery of heart beat did not occur. The isolated hearts of rabbits and rats behaved similarly to the guinea pig heart when exposed to similar amounts of toxic extracts.

Table I shows arterial pressures that were recorded using different perfusion solutions.

Nematocyst toxin extracts did not modify in any significant way the normal action potential of the desheathed nerve used.

TABLE I

Perfusion solution	Pressure in mm Hg
Mammalian Ringer's solution (Krebs)	20
144 mM KCl solution	80
1 μg Adrenaline per ml of mammalian Ringer's solution	120
Toxin extract containing material from 100,000 nematocysts	320

With respect to capillary leakage studies in rat skin, it was noted that injection of histamine caused immediate capillary leakage. The area surrounding the site of injection became very dark blue in color (Fig. 3). Injection of C. fleckeri nematocyst toxin caused the formation of a raised lump. Capil-

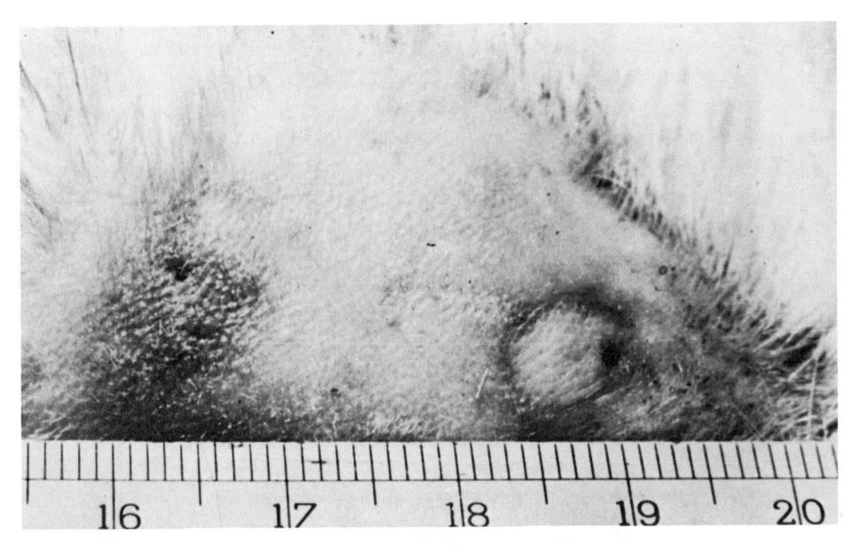

FIG. 3. Photograph of shaved area on back of rat, which had been injected intravenously with 2% Evan's Blue in 0.9% saline. One-fifth of a milliliter of 800 mM histamine was injected subcutaneously on the left-hand side of the shaved area, and 0.1 ml of C. fleckeri nematocyst toxin extract containing material released from 1 x 10^5 nematocysts was injected subcutaneously on the right-hand side. The photograph was taken 10 min after the injections had been completed.

lary leakage occurred almost immediately at the periphery of this lump, which typically measured approximately 6 mm in diameter. Sixteen hours later it was observed that this raised area had become dark blue and that necrosis of cells in the region had occurred. Injection of C. fleckeri extract, followed 5 min later by EDTA, showed the same effect as did the ordinary toxin extracts, indicating that EDTA under the conditions used did not prevent cell necrosis. Injection of similar doses of nematocyst toxin extracts by the ip or iv routes caused no noticeable capillary leakage in the shaved area of skin.

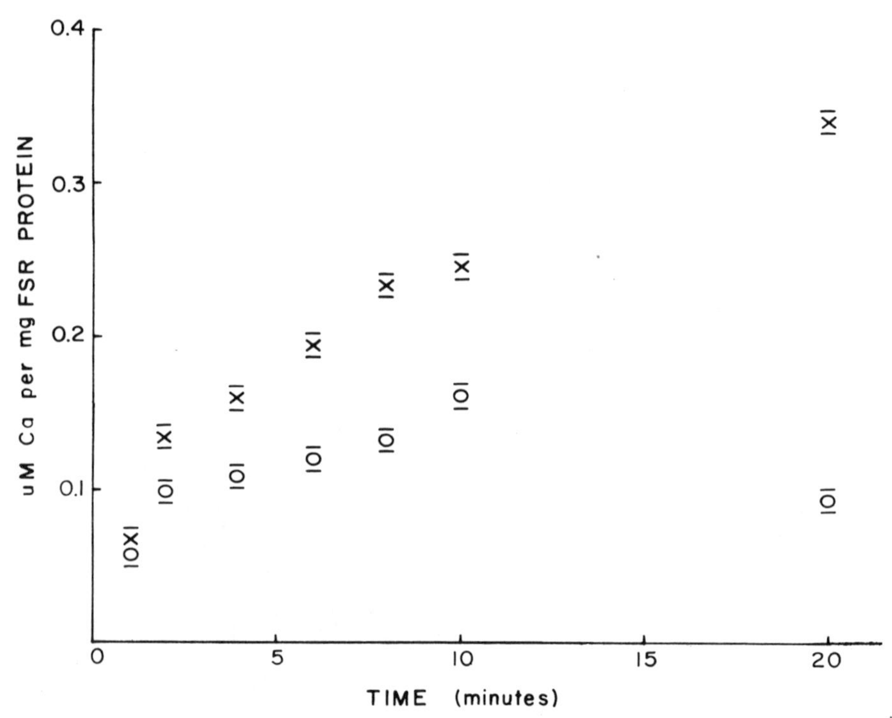

FIG. 4. Graph showing effect of C. fleckeri nematocyst toxin extracts on calcium uptake by fragmented sarcoplasmic reticulum (FSR) extract obtained from mouse striated muscle. X represents value (mean of 4 determinations in each case) obtained for calcium uptake of control FSR extract. O represents value (mean of 6 determinations in each case) obtained for calcium uptake of FSR exposed to nematocyst toxin. Horizontal bars associated with each value indicate limits of deviation from mean. Incubation temperature was 37°C.

Prior treatment of muscles with procaine did not prevent nematocyst toxin extracts from exerting their normal effect on rat diaphragm preparations and rat skeletal muscle preparations; nor did tetrodotoxin block the action of the venom on the rat diaphragm preparation.

After exposure to glycerol Ringer's solution followed by exposure to normal amphibian Ringer's solution, frog sartorius musculature became refractory to electrical stimulation. When challenged with nematocyst toxin, the treated musculature did not contract. However, exposure to 10 mM caffeine caused a sustained contracture of the musculature.

The toxin extract, which was incubated with the fragmented sarcoplasmic reticulum preparation at 37°C for 10 min, affected calcium uptake by the reticulum. The rate of calcium uptake (Fig. 4) was similar to that shown by the control for the first 2 min, but then the preparation, which had been incubated with the nematocyst toxin extract, showed a much slower rate of calcium accumulation. After 20 min the difference in calcium content was 0.25 μM Ca/mg sarcoplasmic protein, which was equivalent to 70% inhibition of calcium uptake compared with the control.

No L-amino acid oxidase activity was exhibited by nematocyst toxin extracts.

IV. DISCUSSION

The mode of preparation of the nematocyst toxin extracts adopted ensured that these extracts were not contaminated with other toxic material normally present in the tentacles (Endean and Noble, 1971). However, it should be noted that material released from all types of nematocyst is present in the extracts.

It has been shown that C. fleckeri nematocyst toxin extracts elicit a sustained contracture of vascular smooth muscle, and this activity would account, to some extent at least, for the rapid onset of a marked hypertension, which is observed to occur in mammals injected with appropriate amounts of nematocyst toxin. However, because the toxin extracts affect other parameters of the cardiovascular system, it is difficult to assess the precise contribution made to the observed hypertension by contracture of vascular smooth musculature.

Isolated hearts of rats and rabbits behave similarly to hearts exposed in situ when challenged with concentrated doses of crude nematocyst toxin. No doubt the observed paralysis of the ventricles in systole is the prime cause of the rapid deaths of animals injected with Chironex nematocyst toxin. The rapidity with which death occurs is a feature of the deaths of human subjects who have received massive stings from Chironex fleckeri. Less concentrated doses of nematocyst toxin do not paralyze the ventricular musculature (Endean et al., 1969) although the activity of the musculature may be impaired.

Capillary permeability studies suggest that initially there is little increase in the permeability to Evan's Blue dye of capillaries exposed to nematocyst toxin, but dye may be released from capillaries after the lapse of some hours. The release of dye may then be due primarily to necrosis of cells in the walls of capillaries. The toxin extracts are certainly capable of causing cell necrosis, but the basis of this activity is obscure. Extracts show no proteolytic activity, phospholipase A activity, or L-amino acid oxidase activity.

The lack of activity of nematocyst extracts on action potentials in desheathed nerve is of interest.

It was shown earlier (Endean and Henderson, 1969) that toxin extracts elicited a contracture in skinned toad muscle fibers and in barnacle muscle fibers where action potentials had been suppressed by exposure to $MnCl_2$ ions. It has now been shown that nematocyst toxin will elicit contractures in mammalian musculature paralyzed by exposure to tetrodotoxin or procaine. Tetrodotoxin is known to interfere with the generation of action potentials by blocking the diffusion of sodium ions into cells (Mosher et al., 1964; Kao, 1966; Russell, 1967), whereas procaine is believed to affect both sodium and potassium ion movements (Shanes et al., 1958; Taylor, 1959). All these results indicate that depolarization of external membranes of muscle fibers may not be essential for the action of the nematocyst toxin. It is of interest to recall that the contracture of musculature elicited by caffeine is likewise independent of depolarization (Axelsson and Thesleff, 1958; Huddart and Abram, 1969). However, the action of caffeine is also independent of the integrity of the transverse tubular system (Gage and Eisenberg, 1967) and does not depend on the influx of extracellular calcium because it can occur in a calcium-free medium (Frank, 1960). In contrast, the integrity of the transverse tubular system has now been shown to be essential for the production of contractures by the nematocyst toxin, and Endean and Henderson (1969) have shown that the contracture elicited by the toxin will not occur in a calcium-free medium.

According to current theory the triads are sites of calcium release during the excitation-contraction coupling cycle

(Langer and Brady, 1963; Winegrad, 1965; Legato and Langer, 1969) and a similar role has been postulated for the dyads of crustacean muscle (Huddart and Oates, 1970). It is also known that an active sequestering of calcium ions by the membranes of the sarcoplasmic reticulum results in relaxation of muscle (Carvalho, 1966, 1968).

Endean and Henderson (1969) have shown that exposure of barnacle muscle fibers to Chironex nematocyst toxin extracts results initially in an influx of calcium ions into the musculature. Apparently this influx is associated with a release of calcium ions into the vicinity of the contractile apparatus resulting in contracture. It is not known whether a similar influx of calcium ions occurs in the case of mammalian musculature exposed to the extracts but this is certainly possible and warrants investigation.

On the other hand, it has now been shown that Chironex nematocyst toxin will inhibit markedly the uptake of calcium ions by the sarcoplasmic reticulum of mammalian musculature, and this could account for the sustained contracture elicited by the toxin in mammalian musculature. In this respect the activity of the nematocyst toxin resembles that of caffeine, which also produces a strong inhibition of the uptake of calcium ions by the sarcoplasmic reticulum (Weber and Herz, 1968; Fuchs, 1969). Possibly the sarcoplasmic reticulum of barnacle musculature is similarly affected by the toxin and this possibility deserves attention.

Eventually the contractures elicited in mammalian and barnacle musculature decline and the musculatures become paralyzed in the relaxed state. Endean and Henderson (1969) have shown that the initial influx of calcium ions elicited by the nematocyst toxin in barnacle musculature is followed by a pronounced efflux. Possibly prolonged exposure to the toxin results in destruction of triads or dyads as occurs in frog and crayfish skeletal muscles, respectively, after prolonged exposure to caffeine (Huddart and Oates, 1970), and this possibility warrants investigation. At any rate, calcium ion movements through muscle fiber membranes are markedly affected by the nematocyst toxin. Possibly, interference with calcium ion movements through cell membranes is also responsible for the dermonecrotic effects produced by the toxin. It would seem probable that, when isolated, the active fraction or fractions of the nematocyst toxin could well become valuable tools in furthering knowledge of excitation contraction mechanisms within muscle and knowledge of calcium transport through membranes generally.

ACKNOWLEDGMENTS

The authors are indebted to A. Hansen of Proserpine, Queensland for collecting the specimens of C. fleckeri used in this study. The work was supported by a grant from the Australian Research Grants Committee.

REFERENCES

Axelsson, J., and Thesleff, S., Acta Physiol. Scand., 44, 55 (1958).
Barnes, J. H., Symposia of London Zoological Society No. 16, 307 (1966).
Carvalho, A. P., J. Cell. Physiol., 67, 73 (1966).
Carvalho, A. P., J. Gen. Physiol., 52, 427 (1968).
Cleland, J. B., and Southcott, R. V., National Health and Medical Research Council Special Report Series No. 12, 1 (1965).
Dimitrov, G. D., and Kankonkar, R. C., Toxicon 5, 213 (1968).
Endean, R., and Henderson, L., Toxicon 7, 303 (1969).
Endean, R., and Noble, M., Toxicon 9, 255 (1971).
Endean, R., Duchemin, C., McColm, D., and Fraser, E. H. Toxicon 6, 179 (1969).
Frank, G. B., J. Physiol., London 151, 518 (1960).
Fuchs, F., Biochim, Biophys. Acta 172, 566 (1969).
Gage, P. W., and Eisenberg, R. S., Science 158, 1702 (1967).
Gunn, J. A., J Physiol, London 46, 506 (1913).
Huddart, H., and Abram, R. G., J. Exp. Zool. 171, 49 (1969).
Huddart H., and Oates, K., Comp. Biochem. Physiol. 36, 677 (1970).
Ikemoto, N., Sreter, F. A., Nakamura, A., and Gergely, J., J. Ultrastruct. Res. 23, 216 (1968).
Kao, C. Y., Pharmacol. Rev. 18, 997 (1966).
Kingston, C. W., and Southcott, R. V., Trans. Roy. Soc. Trop. Med. Hyg. 54, 373 (1960).
Langer, G. A., and Brady, A. J., J. Gen. Physiol. 46, 703 (1963).
Legato, M. J., and Langer, G. A., J. Cell. Biol. 41, 401 (1969).
Mosher, H. S., Fuhrman, F. A., Buchwald, H. D., and Fischer, H. G., Science 144, 1100 (1964).
Russell, F. E., Fed. Proc. 26, 1206 (1967).

Shanes, A. M., Freygang, W. H., Grundfest, H., and Amatniek, E., J. Gen. Physiol. 42, 793 (1958).
Svihovec, J., and Raskova, H., Toxicon 4, 269 (1967).
Taylor, R. E., Amer. J. Physiol. 196, 1071 (1959).
Waugh, W. H., Circ. Res. 11, 264 (1962).
Weber, A. and Herz, R., J. Gen. Physiol. 52, 750 (1968).
Winegrad, S., J. Gen. Physiol. 48, 455 (1965).

Chapter 2

THE BIOLOGICAL ORIGIN AND TRANSMISSION OF CIGUATOXIN

Albert H. Banner

Department of Zoology
Hawaii Institute of Marine Biology
University of Hawaii
Honolulu, Hawaii

The disease ciguatera has been variously defined by many authors (Halstead, 1959, 1967; Russell, 1965; Helfrich et al., 1968; Baslow, 1969; and many others). A good definition must include the concepts that the fish that cause the disease are regionally toxic and usually, if not always, associated with coral reefs through their food chains, and that the disease is primarily neurological, although the early symptoms may include gastrointestinal disorders. The disease cannot be defined by specific fish as carriers, because Halstead (1967) lists over 400 species as ciguatoxic, including both cartilaginous and bony fishes. The disease cannot be well defined by specific symptoms since they are complex and variable (Bagnis, 1968). The disease, as presently defined, cannot even be attributed to a single toxic compound (Yasumoto et al., 1971). It is probable that the definition would not have to include symptoms reported as produced by certain bivalves from Florida in Mc Farren et al., (1965), since this case is unique in the literature and was based on only preliminary chemistry and pharmacology.

The most common and serious symptomatology of the disease has been attributed to a compound isolated and named by Scheuer et al. (1967) as ''ciguatoxin.'' The toxin they reported on came from the moray eel, Gymnothorax javanicus (Bleeker), from Johnston Island. This toxin was reported to be the ''principal toxin'' involved in ciguatera in the Pacific by Banner (1967).

Baslow (1969, p. 206) has suggested that the three principal fish reported on by Scheuer et al. [G. javanicus; Lutjanus bohar (Forsskål), the red snapper; and Carcharinus menisorrah (Muller and Henle) (=amblyrhynchos Bleeker), the grey reef shark] contain different toxins. Scheuer (1970) stated that these fishes ''possess an identical portion responsible for observed physiological activity.'' In a series of studies reported by Rayner and co-workers (1968, 1969, 1970a, 1970b), unique pharmacological properties of ciguatoxin were set forth. Using tests for these unique properties, Rayner was able to state that the acanthurid Ctenochaetus striatus (Quoy and Gaimard) from Tahiti contained a toxin of the same pharmacological activity as did the eels from Johnston Island, 3500 miles distant (Yasumoto et al., 1971). More recent experiments, as yet unpublished, has shown that L. bohar from Palmyra in the Line Islands also has a toxin of the same activity (Rayner, personal communication).

The ciguatoxic fish from the Marquesas and Societies (Bagnis, 1967) to the Ryukyu Archipelago (Hashimoto et al., 1969a,b) and to the Great Barrier Reef (Tonge et al., 1967) cause the same basic neurological symptoms. The scattered reports from the Indian Ocean point to the same conclusions (Halstead, 1967) so it is probable they too have a similar toxin. The toxic moiety may, however, be associated with a range of diferent lipids. For instance, Scheuer has even found a difference in the lipid associates in the flesh and liver toxin of the moray eel from Johnston Island (P. Scheuer, personal communication).

Circumstantial evidence suggests that the same toxic moiety may be found in both the Pacific and the Caribbean. The same families and genera of fishes (such as surgeonfishes, parrotfishes, snappers, groupers, etc.) and even the same species [such as Sphyraena barracuda (Walbaum) and Seriola dumerili (Risso)] cause ciguatera in both areas. These ciguateric fishes occupy the same ecological niches in the coral reef ecosystem in the Caribbean as in the Pacific. The neurological symptoms in both areas are similar (Halstead, 1967, inter alios). Thus until evidence is presented to the contrary, the presence of the same toxin in both areas is the most reasonable scientific presumption. In this review, although most evidence is drawn from the Pacific, the discussion of the ecological phenomenon probably applies as well to the Indian Ocean and the Caribbean Sea.

While this paper concerns itself primarily with the biology of ciguatoxin, a knowledge of recent advances in the chemistry and pharmacology of the toxin is desirable for better comprehension of the problem.

Scheuer et al. (1967) and Scheuer (1970) have reported that a highly purified and apparently homogenous toxic extract from the moray had the empirical formula of $(C_{35}H_{65}NO_8)_n$, but since there were at least three N atoms, it would have a molecular weight of 1500-1800. An independent measurement of the molecular weight gave the same weight. An hydrolysate of the lipoidal toxin contained glycerol and a series of long-chain fatty acids as well as the nitrogenous moiety. It contained no phosphorus as was earlier assumed. It was recoverable from eel flesh at the rate of 5-10 ppm and had the IP toxicity in mice of about 0.5 mg/kg (LD_{99}).

Rayner and associates, studying the pharmacology of the toxin, first showed that the anticholinesterase activity attributed to the toxin by Li (1965a,b) could be observed in vitro but not in vivo. They have shown that the primary activity of the toxin is on excitable membranes, where it causes an ionic imbalance by permitting increased Na^+ permeability. This action is antagonized by tetraodotoxin and the Ca^{2+} ion. The more common acute symptoms associated with ciguatera appear to be consistant with the proposed mechanism of action (see Rayner and co-workers, 1968, 1969, 1970a, 1970b; Setliff et al., 1971).

Mention should be made of other toxins occurring in ciguateric fishes. Banner (1967) suggested that on the basis of chemical and pharmacological evidence more than one toxin might be involved in ciguatera. Hashimoto et al., (1969b) have reported that in addition to what appears to be the true ciguatoxin from the Ryukyu and Amami islands, they found a water-soluble toxin in the grouper, Epinephelus fuscoguttatus (Forsskål). Bagnis (1968), on the basis of observed differences in symptomatology in Tahiti, proposed that the herbivores and carnivores might carry different toxins. Most recently, Yasumoto et al. (1971) reported that C. striatus and Acanthurus lineatus (L.) have at least one toxin in addition to ciguatoxin. The authors also had indications that these acanthurids carried other possible toxins in a different season. Only preliminary work has been done in either chemistry or pharmacology on the associated toxins. In the following discussion, the prime emphasis will be on ciguatoxin.

I. BIOLOGICAL PATTERNS OF CIGUATOXIC FISHES

The disease-causing fishes are limited to the tropics and subtropics and to coral reefs. Notable, however, is that the disease appears to be confined to oceanic islands and does not occur either off the continental masses or the great islands of the western Pacific, such as Indonesia and the Philippines (Banner and Helfrich, 1964). The only major exceptions to this rule are the coastal waters of Florida (Morton and Burklew, 1970) and the Great Barrier Reef of Australia. In the latter area, the ciguatoxic fishes are limited to the coral reefs of the middle and outer reef areas, away from the continental shores (G. Broadbent, personal communication). [It should be noted that these conclusions are not entirely shared by Halstead (1967) in his work.]

The narrow geographic range of ciguateric fishes within an island or an archipelago has been long known and often reported. One of the more interesting comprehensive reports is that of Cooper (1964) on ciguatera in the Gilbert Islands. Her report was based on interviews in Gilbertese with the island residents, who told her of their experience derived from their continuing human bioassay as they searched for protein food. In the recent years after World War II, ciguatera was known on 10 of the 16 low islands and atolls. On the individual atolls the Gilbertese reported only restricted areas as harboring toxic fish, while the same species in the rest of the island waters were safe to eat, as is shown in Fig. 1. Similar discontinuous distributions were reported by Banner and Helfrich (1964), for various archipelagoes, and by Bagnis (1967) from Tahiti and from Hao in the Tuamotus (see below).

Within a toxic area, many species of fish become toxic. Apparently all feed either as herbivores on the benthic flora of the reef or as carnivores on the benthic feeding herbivores. Most food chains are from fish to fish, but at least one carnivore known to be toxic, _Monotoxis grandoculis_ (Forsskål), feeds on benthic herbivorous invertebrates (Randall, 1958). The range of toxic fish extends from the reef sharks to the highly specialized bony fishes such as the plectognaths. (The question as to whether the puffers, family Tetraodontidae, may bear ciguatoxin in addition to tetrodotoxin has not been studied.) The native islanders often state ''all of the fish are toxic,'' but on questioning they restrict the statement to the reef fish that they commonly eat. The reports of ciguatera caused by fish restricted in diet to the high seas are most doubtful. Similarly, the extension of ciguateric symptoms to plankton feeders such as herring and sardines may be questioned, particularly since such plankton feeders carry a toxin of entirely different properties (see Halstead, 1967, under ''Clupeotoxic Fishes'').

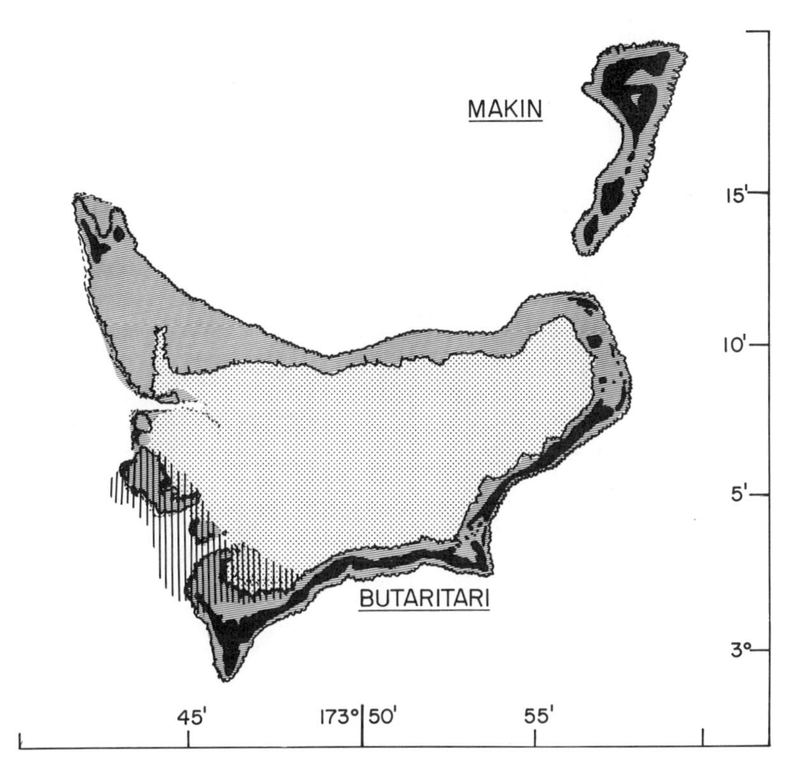

FIG. 1. Map of Butaritari and Makin, Gilbert Islands; vertically hatched area on Butaritari reported to harbor toxic fish; all other areas on both islands with fish safe to eat. (Adapted from Cooper, 1964.)

Within a toxic area, not all fishes are equally toxic (see Halstead's earlier surveys, such as Halstead and Bunker, 1954a,b). Even fish within a population of a single species are not uniformly toxic. From 1964 to 1969, personnel of our program at the University of Hawaii collected 883 specimens of *Gymnothorax javanicus* from Johnston Island. In this program, all eels that weighed less than 10 lb were discarded; the largest individual weighed 43 lb. We found that less than 15% caused four and five reactions in the mongoose test (coma or death in 48 hr; see Banner et al., 1960), the level of toxicity desired for chemical extraction (Fig. 2). A student, R. E. Brock, has applied statistical tests to this sample of eels and found no correlation between size and toxicity, nor any

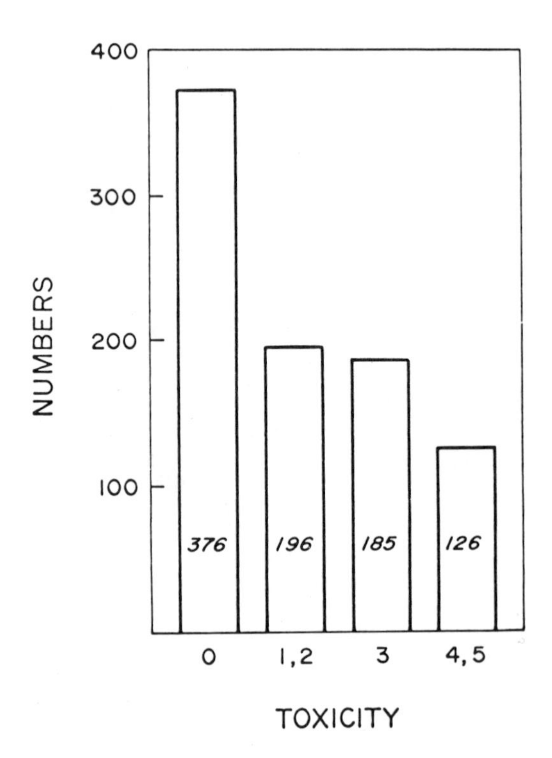

FIG. 2. Toxicity of Gymnothorax javanicus from Johnston Island; 883 eels all over 2.2 kg in weight (average 9.2 kg) caught from 1962 through 1969; toxicity ratings by the mongoose test. As toxicity ratings 1 and 4 appear to be transient, they have been combined with 2 and 5 on this graph.

decline in toxicity during the 5-year period (thesis data, yet unpublished).

In a sample more diverse in size, the toxicity is correlated with size. I have watched Marshallese fishermen on long-toxic Jaluit Atoll discard large specimens of groupers and snappers while retaining smaller fish of the same species for eating. The size differential in toxicity has been confirmed in our large samples of Lutjanus bohar from Palmyra. Figure 3 gives the relationship between toxicity and size in catches of two different years (1959 data from Banner et al., 1963; 1968 data previously unpublished). Helfrich et al. (1968) have analyzed the earlier catches in greater detail and have shown that when the smaller fish are measurably toxic they

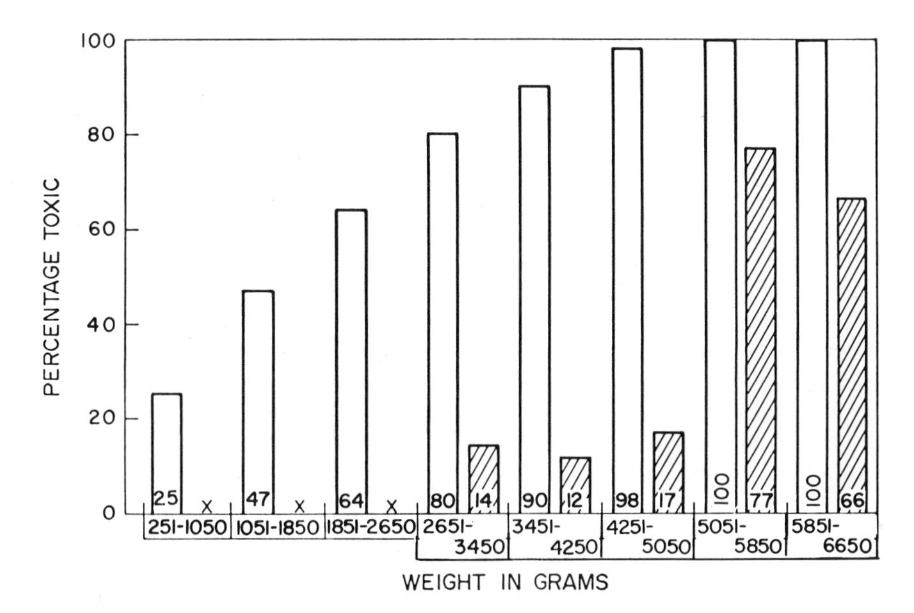

FIG. 3. Percentage of toxic Lutjanus bohar by weight categories, Palmyra, 1959 (solid) and 1968 (hatched), by the mongoose test. 1959 data from Banner et al. (1964), based on 437 specimens; 1968 data based on 168 specimens. Note: in 1968 only one specimen less than 2.6 kg was tested and only 25 in the two highest weight groups.

cause only slight signs of ciguatera in the test mongooses, while a large proportion of the large fish cause marked symptoms and death.

In the islands various time patterns of toxicity exist. Helfrich and Banner (1968) point out that fish from some islands appear to maintain a high or low level of toxicity over many years, even over centuries. The fish in the New Hebrides have been known to be dangerously toxic since Captain Cook was poisoned there in 1774, while in Hawaii, since 1900, outbreaks of ciguatera have been minor and random with the exception of two deaths and one near death in 1964 (Helfrich, 1963; Okihiro et al., 1965). In other islands the number of ciguatoxic fish may show a rapid rise and a slow decline.

The sudden rise in fish toxicity on Hao in the Tuamotu Islands has been documented by Bagnis (1969). When he first surveyed the island in 1964 he was told that there never had been any cases of ciguatera, a disease known to the inhabitants, from local fish. In January, 1965 the French Atomic Energy Commission (CEA) started changes on the atoll to convert it for use as a staging base for their testing program on Mururoa to the south; their work was initiated by the beaching of landing craft from Tahiti and later included piers, dredging, shore-side changes, and other ''improvements'' (Bagnis, personal communication). The work continued for a number of years. The first case of ciguatera was reported from a fish caught at the original French landing site in August, 1966; ciguatoxic fishes then spread to other areas as shown in Fig. 4. In the summer of 1968, Bagnis completed a survey of medical records and a house-to-house survey; the increase in incidence in the population is shown in Fig. 5. (The termination on the graph of cases unreported to medical authorities marks the end of the epidemiological survey; the decline in medically reported cases

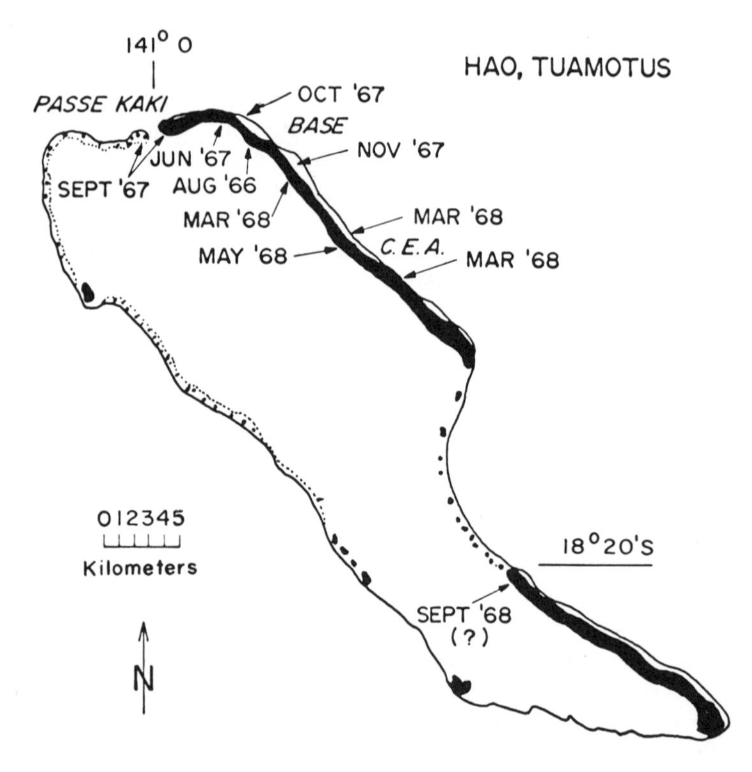

FIG. 4. Hao, Tuamotu Archipelago, showing spread of ciguatera from center of beaching area after August, 1966. (Adapted from Bagnis, 1969.)

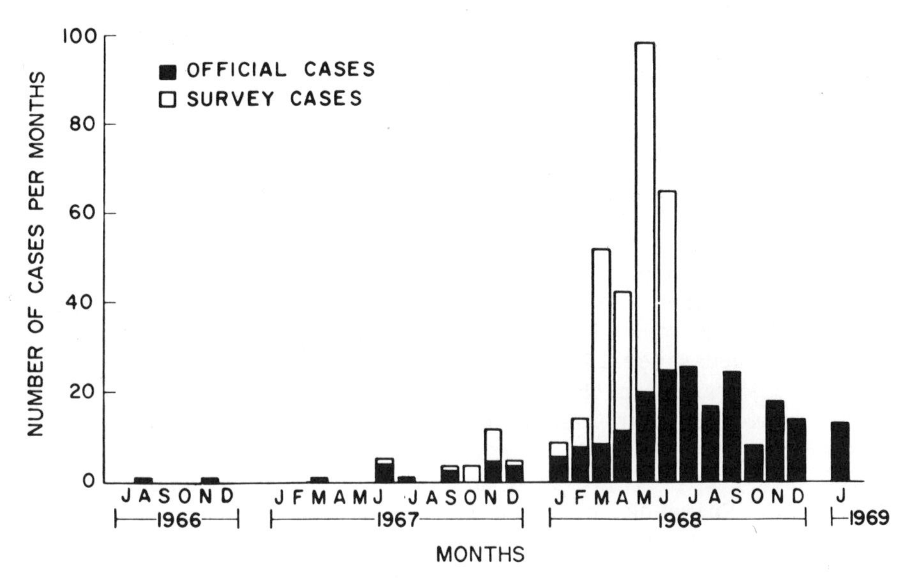

FIG. 5. Monthly incidence of ciguatera on Hao, Tuamotu Archipelago. Ciguatera was unknown to Hao previous to 1966. Solid bar represents the number of official cases reported to medical authorities; the open bar represents the number of additional cases discovered by a house-to-house survey. The epidemiological survey stopped in June, 1968; the official records were available through January, 1969. (Adapted from Bagnis, 1969).

in the later months may reflect a decline in toxicity of the fish, a decline in fish consumption, or a decline in faith in the efficacy of treatment by medical personnel.) The first fishes reported to be toxic on Hao were herbivores, detrital feeders, or scrapers of algae from coral heads (''coralophagous''). It was not until April, 1968, almost 2 years after the initial landings, that the first case of ciguatera from a carnivorous fish was recorded. The four families causing the greatest number of intoxications were the Carangidae, Scaridae, Acanthuridae, and Serranidae. In 1971, Bagnis reported that the fish on the western side of the atoll were still considered to be safe to eat (Bagnis, personal communication).

The decline in toxicity of a fish population has been documented by our laboratory tests on Lutjanus bohar from Palmyra. Helfrich et al. (1968) has reported on testing from

1959 to 1962; Fig. 3 contrasts the 1959 catch with the 1968 catch. As indicated above, a similar decline was not found in the 5-year catch of moray eels from Johnston Island.

II. ORIGIN AND TRANSMISSION OF CIGUATOXIN

Many hypotheses have been advanced to account for the ciguateric type of toxicity in fishes and its spotty distribution. Some of these border on the absurd, such as fishes becoming toxic when the ''coral is in flower'' or fishes becoming toxic from eating the manchineel berry, a superstition that has continued since 1511 (see Halstead, 1967, p. 161). On the atoll of Marakei in the Gilberts the strongly Catholic residents blamed a rapid rise in toxicity to Protestant influences. On Majuro, the Marshallese attributed their rapid rise in toxicity in the early 1950s to the atomic testing program on Bikini and Eniwetok, over 400 miles away, in spite of the fact that fish on Jaluit Atoll, a little over 100 miles away, had been known to be highly toxic for about a century. [Helfrich (1960) showed that there was no correlation between radioactivity in fishes and ciguateric-type toxicity.]

Other hypotheses have been advanced, some superficially plausible (see Halstead, 1967). Most writers have avoided the suggestion that the toxin was a normal endogenous product of the fish, such as tetrodotoxin is in the puffers, for the same species may be toxic in one area and nontoxic in another; moreover, such a unique endogenous product would not be expected to be found in such a broad spectrum of orders. The not infrequent suggestion that the toxin may be a product of postmortal bacterial deterioration has likewise been discarded, for it has been shown that completely fresh fish may cause symptoms. Copper or other heavy metals certainly do not figure directly in the genesis of the toxic moiety, for no metals are in the compound. A number of suggestions in the 1960s by French workers may be viewed with question: Vaillant et al. (1961) suggested that there might be a relationship between ciguatera and eosinophilic meningitis caused by the rat lung worm; Bouder et al. (1962) suggested a similar relationship to type-E botulism; Morice (1965) suggested a causative relationship between the parasitic isopod found in the nostrils of fish and ciguatera; Morelon and Niaussat (1967) recommended that the distribution of organophosphorus insecticides be studied in relation to ciguateric fish; and Chemier (1968), viewing the presence of nickel mines in New Caledonia, reported the successful treatment of 13 patients suffering from ciguatera with a compound that would chelate the nickel ion.

The most widespread hypothesis - now qualifying as a theory - is that the toxin arises in some plant or animal on the coral reef and is transmitted through the food web to the large toxic carnivores. Whereas most writers suggest an alga as the elaborator of the toxin, other forms have been discussed, which include corals, worms, molluscs, and even puffers and other fishes (Halstead, 1967). Hutner and McLaughlin (1958) suggested that the toxin, which they postulated was similar to saxitoxin of paralytic shellfish poisoning, could have arisen from the zooxanthellae in corals. However, the toxin has been shown to be quite unlike saxitoxin in chemistry and pharmacology; moreover, very few fishes eat living coral. The spawning of the palolo worm has been blamed for ciguatera in Fiji especially, but ciguatera occurs where the swarming of the palolo is unknown. De Sylva (1963), after pointing out that the Great Barracuda feeds in part on puffer fishes, suggests the ciguatoxic barracuda may be harboring puffer toxins, but tetrodotoxin has been shown to be of different chemistry and with antagonistic pharmacological action.

The most comprehensive review of the food chain theory is that of Randall (1958). Randall discusses the food habits of toxic fish and finds that they are all tied to benthic life of the coral reef through the food chain. He examines in detail the food habits of the toxic herbivores, especially the acanthurids, and concludes that the originator of the toxin is either a fine benthic alga, such as a cyanophyte or a microbial fungus, protozoan, or bacterium. He suggests that the larger carnivores obtain the toxin by eating the herbivores and that they may actually accumulate the toxin.

Since the publication of Randall's study, most steps have been confirmed. Helfrich and Banner (1963) reported experiments to show that a normally nontoxic omnivore, _Acanthurus xanthopterus_ Cuvier and Valenciennes, could be made lethally toxic (by the mongoose test) by feeding it small amounts of toxic fish flesh daily. They observed no deleterious conditions in the acanthurid from the toxic diet. In a second experiment, Banner et al. (1966) reported that toxic _L. bohar_ from Christmas Island held in ponds in Hawaii for up to 30 months on a nontoxic diet showed no statistically detectable change in toxicity. From these experiments the authors conclude that (1) the toxin may be passed through the food chain without harm to the fishes; (2) the toxin is of exogenous origin, for if it were endogenous the _L. bohar_ would have had an increase in toxicity during the 30 months; (3) the toxin is neither metabolized nor excreted for long periods of time in L. bohar (at the same time the authors warn that this may not be true of all species); and (4) for L. bohar, at least, there can be no marked seasonal differences in toxicity.

The second experiment can account for the toxic eels and sharks on Johnston Island where almost no other species are toxic (Brock et al., 1965). The eels are long-lived fish (Helfrich and Banner, 1968) and either may accumulate the toxin from their prey which might carry it in such small amounts that its presence would be undetectable in laboratory bioassay or may actually be carrying toxin from the outbreak of toxicity in the 1950s (Halstead and Bunker, 1954b).

The attempts to follow the toxin through the food chain to the original elaborator raised the question as to whether a compound recognizable as ciguatoxin or some nontoxic precursor would be found. If the latter is true, the precursor would be altered by the carrier's metabolism to a toxic form. These doubts were settled and the food chain origin and transmission of the toxin were completely confirmed by the work of Yasumoto et al. (1971). The authors studied a large number of specimens of two species of acanthurids, C. striatus and A. lineatus, from toxic zones in Tahiti. They reported that they could extract by the standard method for ciguatoxin a toxin that had not only the reported solubilities of ciguatoxin, but also had the same unique pharmacological activities. They concluded that the toxin from C. striatus was either identical or extremely similar to that of G. javanicus in the toxic moiety. This toxin was found in the flesh, in the visceral organs, and in the stomach contents of these fishes. The presence of the recognizable toxin in the gut contents confirms that toxin is elaborated in a toxic form and in detectable quantities by some organism of the base of the food chain, either an autotroph or a heterotroph.

The study does not indicate the species or even the phylum of the elaborator of the toxin. Ctenochaetus striatus is a detrital feeder and the authors point out that it is unable to break off even ''relatively fine filamentous'' attached alga. Instead, it brushes and sucks up detritus from the bottom. They cite the unpublished work of Walters who found 18 genera of algae in the guts of this species in the same toxic area in Tahiti. They report of a rough fractionation of the pooled stomach contents into three portions: the first primarily of sand, the second primarily of algal fragments, and the third of "unidentifiable particles." It was the third fraction that had the highest yield of ciguatoxin per unit weight.

The question as to whether the elaborator of the toxin is algal or microbial has not been resolved. Dawson et al. (1959) suggested that the toxin might arise from the cyanophyte Lyngbya majescula Gomont, a known toxic form; the rejection of the species as the elaborator by Banner et al. (1960) has been confirmed by the pharmacology of Moikeha and Chu (1971). The possibility of the toxin arising from Schizothrix calcicola (Agardh) Gomont, another bluegreen alga, mentioned by both Cooper (1964) and Banner (1967), has been also rejected, based on preliminary laboratory tests which showed that the toxin was not related either chemically or pharmacologically to ciguatoxin (unpublished). Similarly, Randall (1968) reported on preliminary tests of 19 genera and species of fine algae from Tahiti and the Ryukyus. None of these showed either marked toxicity or did the toxins, when they were present, appear to be related to ciguatoxin.

Yasumoto et al. (1971) suggested that the elaborator is "of small size and low in specific gravity." K. R. Gundersen (personal communication) initiated studies on microorganisms found in the bottom debris and in the gut of C. striatus in Tahiti; in 25 of the 35 mold and yeast isolates that he was able to culture he found some degree of toxicity. However, before even preliminary chemical and pharmacological tests could be performed on the toxins, the project was suspended by lack of funds.

The work of Gundersen pointed out another possibility on the origin of ciguatoxin. Ctenochaetus striatus has an extensive microflora in its gut, many species of which must not be truly pathogenic. This flora must be transmitted through the environment, but at least many of the forms would be adapted to the gut environment and would not be found flourishing in the reef detritus. One of these symbionts could originate the toxin, possibly even as a normal by-product of its metabolism, which would then be absorbed and stored by its host. Similar elaboration of vitamin B_{12} by heterotrophic bacteria is well known.

Although the question of the elaborator of the toxin is yet to be answered, on the basis of logic we can indicate where future research would be the most profitable. First, as it is the simplest and most basic, the study of the gut microflora should be attempted. If these heterotrophs do not produce the toxin in pure culture, then the question arises as to whether to seek first a fine attached alga or an environmental heterotroph. An indication is gathered from the studies of Malardé et al. (1967) on 2798 cases of ciguatera in Tahiti in 1966. Of these, 61.24% were caused by eating C. striatus. Other species of acanthurids contributed 4% or less of all cases. Although the Tahitians tend to eat more C. striatus than other acanthurids and while C. striatus may contain other toxins, the wide difference in percentages must reflect a much higher toxicity in C. striatus than in other acanthurids. The other benthic-feeding surgeonfishes crop fixed algae with fixed cutting teeth; only fishes of the genus Ctenochaetus have flexible brushlike teeth (Randall, 1955). Therefore, to account for the greater toxicity of C. striatus, one must seek those dietary elements not held in common with the fish of lesser toxicity. Since both eat algae, either attached or in broken strands, the importance of that food for our consideration must be reduced. The difference apparently lies in the microbial part of the diet.

If this postulated toxigenic species were in abundance in the detritus, as it would have to be to cause the high percentage of toxic fish that have been reported, wave and current action would carry it from the bottom to rest on, and contaminate, normally growing algae, both in normal bottom stands and on surfaces of dead coral boulders. Thus it could be eaten by the other acanthurids or scraped off the coral by the scarids and enter the food web. This would give the lesser toxicity to the nondetrital feeding fishes, as has been reported.

III. THE ECOLOGY OF OUTBREAKS

As with the speculation on the original elaborator of ciguatoxin, many suggestions have been made about the changes in reef ecology that cause an area to become suddenly toxic, as Bagnis has recorded for Hao. It is true that a combination of many factors, varying from island to island, could cause the appearance of the toxin in the ecosystem. However, at the present state of knowledge it would be simpler and perhaps more scientific to attempt to set up a uniform hypothesis that would cover all known outbreaks, if possible.

Under the criterion of uniformity, several hypotheses can be discarded. On an historical basis, the suggestion of Morelon and Niaussat that modern insecticides have changed the ecology as well as the more sophisticated approaches to atomic testing programs, which suggest the changes in the ecology of distant islands, can be ruled out. Several have suggested that although heavy metals, as from mines or shipwrecks, have not entered into the composition of the toxin, their presence as ions has altered the environment to encourage the change in the floral balance and the production of the toxin. Since islands without metals in their soils and without major shipwrecks for decades, such as some of the coral atolls in the Gilberts and Phoenix groups, are found to be toxic, the suggestion does not appear to be valid.

Probably the most widely accepted hypothesis held today, that of Randall (1958), is both difficult to accept or disprove. In his development of the food-chain theory, Randall suggested an ecological cause for the onset or increase in ciguatoxic fish. He points out that a ''new surface'' on a coral reef may be caused by a number of factors wherein the living cover of coral or other biota is destroyed, leaving a bare surface for recolonization and ecological succession. He lists many factors that could cause extensive new surfaces: the anchoring (or dragging anchor) of many ships, as during World War II in the Line Islands; the flooding of reef areas by fresh water drainage during torrential storms; the scour of storm-driven waves; the collapse of steep reef fronts under hurricanes, earthquakes, or tsunamis; and the rubble dropped by mining operations or dredging.

Randall even mentions the denuding of surfaces by grazing animals, especially the parrot fishes; he states ''in this case the effect is generally too diffuse . . but . . each isolated little scraped area becomes a focus for settling forms.'' Although his other projections of causes of new surfaces are catastrophic enough to account for unique changes in ecology, this suggestion must be discarded since all coral reefs, toxic or not, have such microdenudation continuing at all times. Such nibbling cannot be considered as a unique trigger action.

On these newly bare surfaces, bacteria, plants, and animals would appear in ecological succession. Randall postulates that an alga, possibly a blue green – group known to contain many toxic species – would be early in the succession. One of these would be toxigenic, fine and soft, and cropped even by the herbivores with fine teeth. Thus, he reasoned, an increase in new surfaces would permit the introduction of toxins into the food chains.

He warns that the exposure of new surfaces would not necessarily cause an area to become toxic and cites submarine changes in Hawaii both from wartime activity and from the numerous lava flows that plunged into the sea. He states:

''Clearly other environmental conditions must be satisfied.'' What these may be, he does not discuss.

Many outbreaks of ciguatera can be cited to support his hypothesis; unfortunately, many outbreaks, or lack of outbreaks, can be cited that do not support it.

As indicated, Randall cites a number of areas where bottom changes have preceded an increase in toxicity of the fish. Since his paper was written, more of these changes have occurred. In the Line Islands, Washington Island was the only island untouched by the war and the only island where ciguatera was unknown. In December, 1964 a freighter ran aground on the island and started to break up; by September, 1965 a correspondent reported that ''all of the Lutjanus, Caranx, Scarus, etc., have gone violently toxic....'' (Helfrich et al., 1968). Cooper (1964) states that in the Gilberts ''All of these islanders agree with Randall....'' and pointed to dumped war-surplus materials, shipwrecks, etc., as the cause of the toxicity on their islands. However, the connection between the toxicity and the event to which it was attributed became somewhat tenuous in some cases, such as at Abemama where the vessel was wrecked in 1917 or at Tabiteuea where one was wrecked in 1919 and the toxicity continued at least until 1961.

Probably the best correlation between ''new surfaces'' and ciguatera is in the case of Hao, already cited. Here the atoll was nontoxic until the French started their changes in the marine environment. According to Bagnis (personal communication), the spread of ciguatera to various parts of the atoll followed by 1.5 to 2 years the local changes in the reef environment. This could be viewed as an excellent confirmation of Randall's hypothesis.

However, major changes in marine environments elsewhere in the Pacific, even in areas that had concurrent or previous histories of ciguatera, have not produced outbreaks. Few atolls had more gross disruption than did Eniwetok during the atomic testing program, yet while other atolls in the Marshalls had severe outbreaks of ciguatera (Bartsch and McFarren, 1962), Eniwetok was reported to have produced no ciguatoxic fishes (Helfrich, 1960). In 1962, the Gilbert and Ellice Colony had reef passages blasted in nine islands of the Ellice, Gilbert and Phoenix; according to the colony medical officer, Dr. Peter Matthews (personal communication), no increase in toxicity was found. Johnston Island, like the Gilberts, harbored many toxic fishes in the early 1950s (Halstead and Bunker, 1954b), yet after extensive dredging in the 1960s, no fish were found to be toxic other than a few large carnivores (Brock et al., 1965, 1966).

Even vast changes by natural forces did not increase toxicity. The Samoas were hit by a devastating typhoon in 1966, Saipan in 1968, yet no increase in toxicity was noted in the reef fishes. Cooper (1966) records the ruin of the biota of the Mbau Waters by the temporary shift of the mouth of the Rewa River during a typhoon on Viti Levu, Fiji; again there was no increase in toxicity of the fish in the lagoon (Cooper, personal communication). Even the destruction of the living coral on the reefs of Guam by the starfish _Acanthaster_ (Chesher, 1969) resulted in no increase in ciguatoxic fish according to the Fisheries Officer (I.I.Ikehara, personal communication).

On the other hand, the rapid increase in toxicity of the Tahitian fishes in the mid-1960s paralleled extensive changes in Papeete Harbor, yet the most toxic areas were not at or near the harbor, but on undisturbed reefs half-way around the island (Bagnis, 1967; Malardé et al., 1967). At the same time ciguatera was increasing in the waters of Tahiti, it was also increasing in Bora Bora, almost 150 nautical miles away where no changes in the marine environment were made (Bagnis, 1967).

Randall, himself, reported that he had put out ''new surfaces'' of plywood, asbestos board, and concrete in the toxic Papara/Mataeia districts of Tahiti but could find none of the attached alga either toxic enough or with the right toxins to account for the ciguatera in the fishes of the adjacent reef (1968). He suggested that the algae he had studied in a preliminary fashion might carry the toxin at such a low level that it would be undetectable by laboratory tests. This suggestion can be ruled out on the basis of the later studies of Yasumoto et al. (1971) who showed that toxin can be detected in dietary sources. Randall also suggested that since the tests were run during the southern winter, the production of the toxin might be seasonal and be found during the southern summer. It is true that the main sample of _Ctenochaetus_ studied by Yasumoto et al. was made in late January.

Unfortunately there is little to offer either to supplement or to replace Randall's hypothesis. Helfrich and Banner (1968) point out that a biologically stable toxin could be recycled in an environment, and that the fishing practice of discarding potentially toxic fish or their viscera would tend to recycle further the toxin. This would only contribute to the maintenance of the toxicity in an area, but not cause its increase.

With pollution and eutrophication so much in the public consideration, ''enrichment'' might be considered as an ecological cause of an increase in production of ciguatoxin. The increase of toxicity at Hao, for example, could as well be attributed to disposal of the wastes of the increased human population as to the bottom changes. But there are many islands where the waters were not so ''enriched'' yet the fish were still toxic. Cooper (1964) speaks of the toxicity of the fish on Sydney Atoll in the Phoenix, an island that held only a small settlement of Gilbertese. The Marshallese from Bikini were moved first from Bikini to Rongerik, previously without permanent inhabitants; one of their objections to their newly assigned home was that the fish were too toxic (Mason, 1950).

Another hypothesis that could be advanced is that a toxigenic form, microbial or algal, is introduced to a new island ecosystem where it would spread widely and then subside. The toxigenic form might be a different species or merely a mutant of a widespread but normally nontoxic species. Again at Hao, a ship's bottom fresh out of Tahiti, which was suffering from widespread toxicity at the time of the buildup on Hao, could have carried the toxic species. Similarly, some of the Gilbertese on Marakei (possibly the Protestants) attribute the onset of ciguatera to a United States Navy landing craft carrying building supplies from Tarawa, where ciguatera was then common (Cooper, 1964). The same author states ''it would appear that shipping may in some way be associated with the spread of toxicity.''

However, if the toxin arises from a form living freely on the reef, it is probable that the changes causing the rise in number of ciguatoxic fish are far more subtle than those considered. It seems most plausible to suggest that the elaborator is widespread but not common in most reef environments. Supporting this view is the fact that no marine animal yet tested by feeding toxic fish reacts to the toxin (Banner et al., 1960); marine crabs, shrimp and lobster do not react, but freshwater crayfish do. This would suggest that the marine forms had evolved a method of detoxification of a toxin long found and widespread in their environment. This would account for the sporadic cases as are found in islands like Hawaii. A slight change in temperature, a slight increase in nitrates, and an overgrowth of plants causing more detritus, even the disruptions cited by Randall, could shift the delicate balance in the intensely competitive ecosystem, favoring the growth of the toxigenic form. In some cases as in the New Hebrides or on

Jaluit in the Marshalls, the favorable conditions would continue and the fish would remain toxic. In others, the changed conditions would shift back to the original and slowly the biologically stable toxin would pass from the environment.

With the postulation of a toxigenic member of the intestinal flora, it is easier to account for an outbreak such as that at Hao or Palmyra. Again the toxigenic species would be passively introduced to an area where it was unknown, as on a ship's bottom. It might live as a bottom heterotroph or rest as a spore. In the first case it could actively spread in the new environment, in the second it would live in the environment of the gut and be spread through spores in the faeces. In either case, the species would flourish in the intestinal flora of one or many herbivores. These then would absorb the toxin and spread it through the food web. In the new environment, the toxigenic species would first thrive and then slowly be dominated by competitive species as has often been recorded for introduced species into a previously stable ecosystem, both terrestrial and marine. In some conditions it would continue to be a major factor in the biota.

Any coherent and plausible hypothesis on the ecology of ciguatoxin would have to take into consideration the lack of the toxin from the waters of continental masses and of the great islands. None of the hypotheses thus far advanced, including those above, have so done.

IV. CONCLUSIONS

The known facts on the distribution of ciguatoxin in the marine environment indicate that the original elaborator of the toxin is either a fine alga or a microbial heterotroph; the heterotroph appears to be a more logical choice as the source. This heterotroph could either exist freely in the coral reef detritus or be part of the gut flora of some of the marine herbivores; the latter possibility should not be overlooked. No hypothesis yet advanced for the sudden rise in ciguatoxic fish in a reef ecosystem meets with all the reported facts. Again, the infestation of the gut of herbivores by heterotrophs appears to be plausible. Any hypothesis on the ecology of ciguatoxin cannot be substantiated or denied until the species that elaborates the toxin is isolated and studied.

ACKNOWLEDGMENT

This work was supported in part by Public Health Service Grant FD-00396-01, ''Ciguatera in the Pacific.''

REFERENCES

Bagnis, R., Rev. Intern. Oceanogr. Med. 6-7, 89 (1967).
Bagnis, R., Hawaii Med. J. 28(1), 25 (1968).
Bagnis, R., Rev. Corps. Sante Armees 10(6), 783 (1969).
Banner, A. H., In ''Animal Toxins'' (F. E. Russell and P. R. Saunders, eds.), pp. 157-165. Pergamon Press, New York (1967).
Banner, A. H. and Helfrich, P., Hawaii Institute of Marine Biology Technical Report No. 3 (1964).
Banner, A. H., Helfrich P., and Piyakarnchana, T., Copeia (2), 297 (1966).
Banner, A. H., Helfrich, P., Scheuer, P., and Yoshida, T., Proc. Gulf and Carib. Fish. Inst. 16th annual 1963, 87 (1964).
Banner, A. H., Shaw, S. W., Alender, C. B., and Helfrich, P., South Pacific Commission Technical Paper No. 141, 17 (1963).
Banner, A. H., Scheuer, P. J., Sasaki, S., Helfrich, P. and Alender, C. B., Ann. N. Y. Acad. Sci. 90, 770 (1960).
Bartsch, A. F., and McFarren, E. F., Pacific Sci. 16, 42 (1962).
Baslow, M. H., ''Marine Pharmacology.'' The Williams and Wilkins Co., Baltimore, Maryland, 1969.
Bouder, H., Cavallo, A., and Bouder, M. J., Bull. Inst. Oceanogr. 59(1240), 66 (1962).
Brock, V. E., Jones, R. S., and Helfrich, P., Hawaii Institute of Marine Biology Technical Report No. 5, 90 (1965).
Brock, V. E., van Heukelem, and Helfrich, P., Hawaii Institute of Marine Biology Technical Report No. 11 (1966).
Chemier, G., South Pacific Commission Seminar on Ichthyosarcho-toxism (SPC/ICHT/WP-9), 2 (1968).
Chesher, R. H., Science 165, 280 (1969).
Cooper, M. J., Pacific Sci. 18(4), 44 (1964).
Cooper, M. J., Pacific Sci. 20(1), 137 (1966).
Dawson, E. Y., Aleem, A. A., and Halstead, B. W., Allan Hancock Foundation Publication, Vol. 17, pp. 1-39. University of Southern California Press, Los Angeles, 1959.
DeSylva, D. P., Stud. Trop. Oceanogr. Miami, Vol. 1, 179 pp., October, 1963. Institute of Marine Sciences, University of Miami Press.

Halstead, B. W., ''Dangerous Marine Animals.'' Cornell Maritime Press, Maryland, 1959.

Halstead, B. W., ''Poisonous and Venomous Marine Animals.'' Vol. II, 1070 pp. U. S. Government Printing Office, Washington, D. C., 1967.

Halstead, B. W., and Bunker, N. C., Copeia (1), 1 (1954a).

Halstead, B. W., and Bunker, N. C., Zoologica 39(2), 61 (1954b).

Hashimoto, Y., Konosu, S., Yasumoto, T., and Kamiya, H., Bull. Japan. Soc., Sci., Fisheries 35(3), 316 (1969a).

Hashimoto, Y., Yasumoto, T., Kamiya, H., and Yoshida, T., Bull. Japan. Soc. Sci. Fisheries 35(3), 327 (1969b).

Helfrich, P., U.S. AEC, Tech. Inf. Serv. Publ. TID-5748 (1960).

Helfrich, P., Hawaii Med. J. 22, 361 (1963).

Helfrich, P., and Banner, A.H., Nature 197, 1025 (1963).

Helfrich, P., and Banner, A.H., Bernice P. Bishop Museum, Occasional Papers 23(14), 371 (1968).

Helfrich, P., Piyakarnchana, T., and Miles, P.S., Bernice P. Bishop Museum, Occasional Papers 23(14), 305 (1968).

Hutner, S. H., and McLaughlin, J. A., Sci. Am. 199, 92 (1958).

Li, K. M., Science 147, 1580 (1965a).

Li, K. M., Hawaii Med. J. 24, 358 (1965b).

Malardé, L., Bagnis, R., Tapu, J., Bennett, J., and Nanai, F., Institute Recherches Medicale Polynesie Francaise, Technical Report No. 1, August, 1967.

Mason, L. E., Human Organization 9(1), 5 (1950).

McFarren, E. F., Tanabe, H., Silva, F. J., Wilson, W. B., Campbell, J. C., and Lewis, K. H., Toxicon 3, 111 (1965).

Moikeha, S. N., and Chu, G. W., J. Phycology 7(1), 8 (1971).

Morelon, R., and Niaussat, P., Cahiers du Pacifique No. 10 (1967).

Morice, J., Revue des Travaux de L'Institute des Peches Maritimes 29(1), 4 (1965).

Morton, R. A., and Burklew, M. A., Toxicon 8, 317 (1970).

Okihiro, M. M., Keenan, J. P., and Ivy, A. C., Hawaii Med. J. 24, 354 (1965).

Randall, J. E., Zoologica 40, 149 (1955).

Randall, J. E., Bull. Marine Sci. Gulf Caribbean 8(3),236 (1958).

Randall, J. E., South Pacific Commission Seminar on Ichthyosarcotoxism, August, 1968.

Rayner, M. D., The Pharmacologist 12, abstract 546 (1970a).

Rayner, M. D., In ''Proceedings of Food/Drugs From the Sea Conference 1969'' (H. W. Youngken, ed.), pp. 345-350. Marine Technology Society, Washington, D. C., 1970b.

Rayner, M. D., Baslow, M. H., and Kosaki, T. I. J. Fisheries Res. Board Canada 26, 2208 (1969).

Rayner, M. D., Kosaki, T. I., and Fellmeth, E. L., Science 160, 70 (1968).

Russell, F. E., In ''Advances in Marine Biology'' (F. S. Russell, ed.), pp. 255-389. Academic Press, London, 1965.

Scheuer, P. J., Advan. Food Res. 18, 141 (1970).

Scheuer, P. J., Takahashi, W., Tsutsumi, J., and Yoshida, T., Science 155, 1267 (1967).

Setliff, J. A., Rayner, M. D., and Hong, S. K., Toxico. Appl. Pharmacol. 18, 676 (1971).

Tonge, J. I., Battey, Y., and Forbes, J. J., Med. J. Australia 2, 1088 (1967).

Vaillant, A., Peyrin, A., Cavallo, A., and Bordes, F. P., Bull. Soc. Pathol. Exotique 54(5), 1075 (1961).

Yasumoto, T., Hashimoto, Y., Bagnis, R., Randall, J. E., and Banner, A. H., Bull. Japan. Soc. Sci. Fisheries 37(8), 724 (1971).

Note added in proof: In a recent informal discussion with the author, Dr. Raymond Bagnis of the Institut Recherches Medicales Louis Malarde of Papeete, Tahiti, described new outbreaks of ciguatera in French Polynesia, especially in the Tuamotus. He is of the opinion that these so closely follow definite disruptions of the reefs that he could safely predict the occurrence of ciguatoxin in a coral reef ecosystem following extensive dredging or other submarine changes. He has not yet published upon these outbreaks.

Chapter 3

MODES OF ACTION AND IDENTITIES OF PROTEIN CONSTITUENTS IN SEA URCHIN TOXIN

George A. Feigen and Lahlou Hadji

Department of Physiology, School of Medicine
Stanford University
Stanford, California

John E. Cushing

Department of Biological Sciences
University of California at Santa Barbara
Santa Barbara, California

I. INTRODUCTION

Man's traditional concern with animal venoms has changed over the past century from a practical study of measures to insure his own survival to the use of venoms as highly specific physiological and biochemical probes. In many instances venoms have played a determining role in the elucidation of critical steps in blood coagulation, in the action of complement, and in the analysis of certain mechanisms important to the understanding of immediate hypersensitivity in man, such as the discovery of slow-reacting substance by Kellaway and Trethewie (1940) and the mode of formation of bradykinin by Rocha e Silva et al (1949).

In spite of their widespread use as specific reagents, information about most animal venoms is still largely fractional, and a general pattern of the design of venom complexes has emerged only in the case of the terrestrial snake. Most snake venoms appear to contain ensembles of proteins, some of which are clearly digestive enzymes whereas others, such as specific neurotoxins, are not. Since they may also contain activators for these enzymes as well as inhibitors for certain enzymes of the recipient, it is evident, as Tu (1971) has pointed out, that the toxicity, in a broad sense, is due to the enzymes as well as to the nonenzymatic proteins. The fact that many snake venoms contain phospholipase A, proteases of various specificities, nucleases, phosphodiesterase, cholinesterases, hyaluronidases, and many other enzymatic proteins, distributes their point of attack to many targets simultaneously, and it then becomes a formidable problem to decide which are the central and which the peripheral agents responsible for the specific envenomation. Some snake venoms are hemolytic and many of them release histamine and other autopharmacological agents, but it is generally conceded that neither hemolysis nor histamine release is the major cause of lethality, although the hypotension and increased vascular permeability can enhance the actions of the other venom components.

To simplify the problem, our aim has been to provide a detailed description of the chemistry and mode of action of a relatively simple venom complex present in the sea urchin Tripneustes gratilla (Linnaeus). This urchin is a common inhabitant of the littoral of the Indo-Pacific. It has on its test globiferous pedicellariae which contain a venom that is highly lethal when injected into mice, rabbits, crabs, lobsters and sipunculid worms. The stings are painful to man, and the local reactions can be aggravated by generalized hypersensitivity in persons repeatedly exposed.

In many respects the actions we have discovered thus far have shown that sea urchin toxin (SUT) may be only qualitatively, but not functionally, different from venoms of higher forms. Similar to snake venoms, SUT has a multiplicity of effects: It has hemolytic, cardiotoxic, and neurotoxic properties; it releases histamine and other physiologically active substances from various visceral tissues; and it can form bradykinin from appropriate substrates in mammalian plasma and destroy it once it is formed. These actions are exerted by an ensemble of six to nine proteins present in the crude extract.

The purpose of this report is to describe in a general way the preparation, isolation, characterization, and pharmacological action of some of these materials and to give a detailed account of the biochemical behavior of the component responsible for the formation of bradykinin.

II. PREPARATION

A. Collection and Processing

Specimens were obtained from three general locations around the island of Oahu. Collections were made from the reefs and shoals in Kaneohe Bay, from the seaward side of the main reef in the Waikiki area, approximately 0.5 to 1.5 miles off-shore, and, last, from an area west of the Ilikai Harbor, approximately 0.5 miles seaward from Ala Moana Park. Pedicellariae were obtained by washing the specimens with a high pressure seawater spray and screening the washings through a coarse and then a fine sieve. As the pedicellariae are removed and concentrated by this procedure, they agglutinate and can then be separated from sand and other detritus by gentle elutriation. The elutriation is repeated several times and the final pack of pedicellariae is centrifuged, placed in tared polyethylene bottles, and frozen at $-20^{o}C$.

B. Fractionation and Characterization

The separation of active materials from the crude seawater extracts was accomplished in three general ways: fractional precipitation in the presence of ammonium sulfate, gel filtration on Sephadex G-200, and chromatography on hydroxylapatite. Although at different times during the development of the work, studies were made for various purposes on the products obtained independently by each of these methods, the procedure currently used includes the initial salt precipitation with the two types of chromatography. The details of this scheme are given below.

C. Starting Material

The frozen packs of pedicellariae are homogenized with four volumes of distilled water for 5 min and the homogenate is centrifuged at 10,000 g. The supernate is removed and the sediment is reextracted twice in the same way.

TABLE I

Yield of Solids from Sea Urchins by Seawater Extraction

Collection (specimens)	Total weight protein (g)	Total N per animal (mg)	Protein N per animal (mg)
Sept. 1967 (2008)	2.31	0.273	0.185
Sept. 1968 (1906)	1.62	0.194	0.136

TABLE II

Properties of SUT Fractionated with Ammonium Sulfate

	Analytical properties [a]					Pharmacological properties		
	Whole preparation		Soluble fraction [b]					
Prep. 64	mg N/mg dry toxin	Prot. N % [c]	Total N mg/ml extract	Prot. N % [c]	OD_{278}	No. Mice	LD_{50} Prot./20 gm mouse (x 10^{-4} mg)	LD_{50}/mg Prot.
Parent	0.0655	70.83	0.0541	85.77	1.067	40	11.06	903
33% SAS	0.1187	21.40	0.0766	33.16	1.301	27	117.50	85
65%	0.1260	74.76	0.1166	80.79	1.694	40	14.50	690
100%	0.1609	33.75	0.1609	33.75	1.031	38	55.50	180

[a] Nitrogen content per milligram of SUT.
[b] Extraction of 1 mg dry material with 1 ml 1% NaCl.
[c] Precipitate N/Total N.

The proteins are brought down in the presence of saturated ammonium sulfate; the precipitate is separated by centrifugation. Storage in the presence of 0.3 M NaCl significantly decreased the loss of potency of the extract over that experienced with lyophilization. The amount of crude extract obtained in the lyophilized preparations is given in Table I and the analytical and toxicological properties of the various fractions are shown in Table II. The stability of the frozen materials, assayed periodically over 292 days, can be assessed from the bioassay data shown in Table III.

TABLE III

Effect of Storage Time on the Potency of Crude Pedicellarial Toxin (SUT-1967)

Time of LD_{50} Determination (days)	LD_{50} (mg N x 10^{-4})	LD_{50} (x 10^3) mg N [a]
0	1.29	7.75
77	1.15	8.60
105	1.75	5.71
139	1.49	6.71
167	1.30	7.75
194	1.63	6.15
230	1.30	7.75
261	1.70	5.88
292	1.49	6.71

[a] Mean LD_{50}/mg N = 7.00 $\pm$ 0.98 (x 10^3) as estimated on TCA precipitates.

D. Chromatography

Further purification can be accomplished by the use of Sephadex or hydroxylapatite as filtration beds. Since the levels of the proteins as well as the specific lethal toxicity appear to be concentrated in Fraction II of the ammonium sulfate preparations, that fraction may be used as the starting material, but since we were also interested in other properties, it was convenient to fractionate the whole starting material.

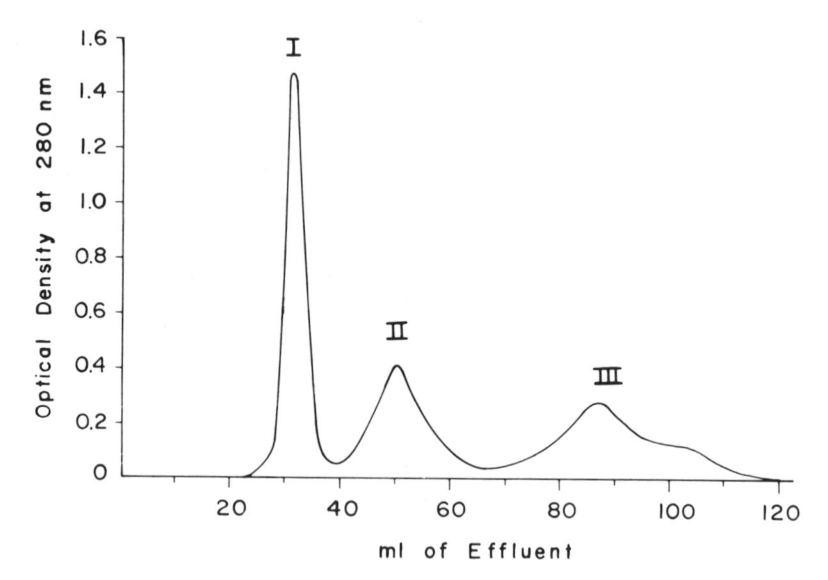

FIG. 1. Elution diagram of SUT(65) in phosphate-buffered 0.25 M NaCl at pH 7. Optical density at 280 nm.

TABLE IV

Precipitable Nitrogen Content, Sedimentation Coefficients, and Lethal Toxicities of SUT Fractions

Preparation		% TCA Precipitable		LD_{50}
Type	Fraction no.	Nitrogen	S_{20}	(mg total N)
Crude		67.6	20 – 6 – 2.1	5.24×10^3
Sephadex G-200	I	40.4	20, 6.4	1.85×10^3
	II	62.0	5.7	1.24×10^4
	III	51.5	2.0	7.3×10^2
Hydroxylapatite	I	53.1	2.2	0
	II	66.1	6.4	1.0×10^4
	III	91	6.4	6.0×10^3
	Intermed.	100	6.5	2.2×10^3
	IV	100	– [a]	5.5×10^3
	V	100	1.8	0
	VI	100	– [a]	0

[a] Insufficient material.

1. Sephadex

Percolation of the extracts obtained from 65 animals on Sephadex G-200 resolved the material into three peaks, as shown in Fig. 1. Table IV shows an efficient concentration of biological activity in Fraction II and suggests that the lethal potency is associated with molecules having a sedimentation constant (S_{20}) of 5.7.

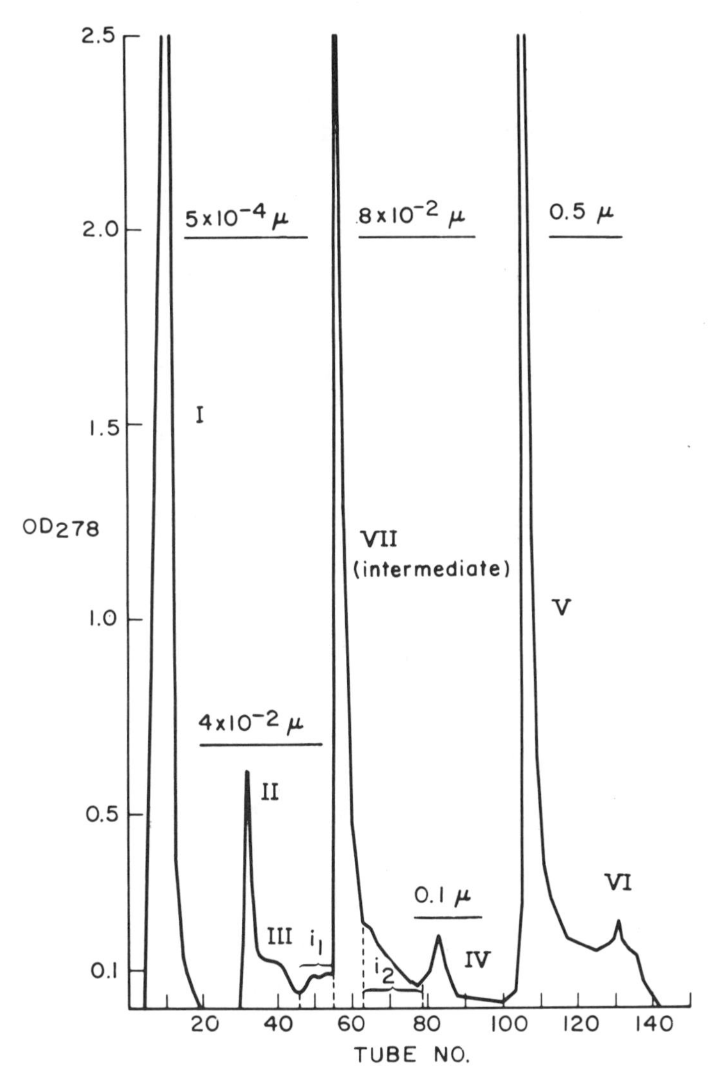

FIG. 2. Fractionation of crude toxin (SUT-67) on hydroxylapatite. M, ionic strength of phosphate buffer at pH 6.8.

2. Hydroxylapatite

The efficiency of separation on hydroxylapatite is shown by the effluogram in Fig. 2. In this experiment a sample of starting material containing 3.2 mg SUT-N was applied to the column in phosphate buffer at pH 6.8. The material was eluted by stepwise increments of ionic strength (Γ/2) ranging from 0.0005 to 0.5. This method of separation showed a greater resolution than Sephadex chromatography, but in both cases the lethal activity appeared to be associated with molecules having approximately the same values. It is also significant that the most potent fractions (Seph II and HOA II) had comparable activities and comparable proportions of protein-N.

E. Factors Affecting Lethal Potency

The crude preparations were shown to be sensitive to repeated freezing and thawing. They were not affected by pH in the range 4.3 to 10.6 when kept in solution at 0°C for 24 hrs. The potency was quickly degraded at temperatures above 40°C. The critical point for a 20-min inactivation appears to lie between 42.5° and 45°C. Complete inactivation at 45°C was observed after 15 min and at 47.5°C after 5 min of incubation.

III. PHARMACOLOGICAL EXPERIMENTS ON ISOLATED MAMMALIAN TISSUES

The object of the pharmacological studies was to determine whether the toxin preparations contained substances capable of releasing histamine and other pharmacologically active agents from the isolated tissues of guinea pigs, rats, and mice that could contribute to the overall lethality of the toxin. Accordingly, studies were first made on the direct effects of toxin as a means of assessing the median effective dose (MED), and, subsequently, pharmacological and chemical analyses were carried out on the dialyzed fluids released during the attack of various toxin preparations on isolated portions of guinea pig ileum, rat colon and lung, and guinea pig heart.

A. Standardization and Bioassay

The response to the direct effects of toxin and to the materials released from toxin-treated tissues were quantitated on the isolated guinea pig ileum. Each gut was first standardized at 37°C by recording its reactions to a range of concentrations of histamine, serotonin or bradykinin as required in the experiment. Immediately following each response the bath was drained and the gut washed three times. A period of 4 min was allowed for equilibration between each challenge, and the testing or calibration procedure was repeated until a consistent response was obtained. When the gut strip had been thus calibrated with the standard compound and the overall maximum contraction had been recorded, it was challenged with a suitable dose of toxin or with several dilutions of bath fluid containing material released from the tissue-toxin reaction mixture. At the end of the assay, each gut strip was recalibrated with the standard compound.

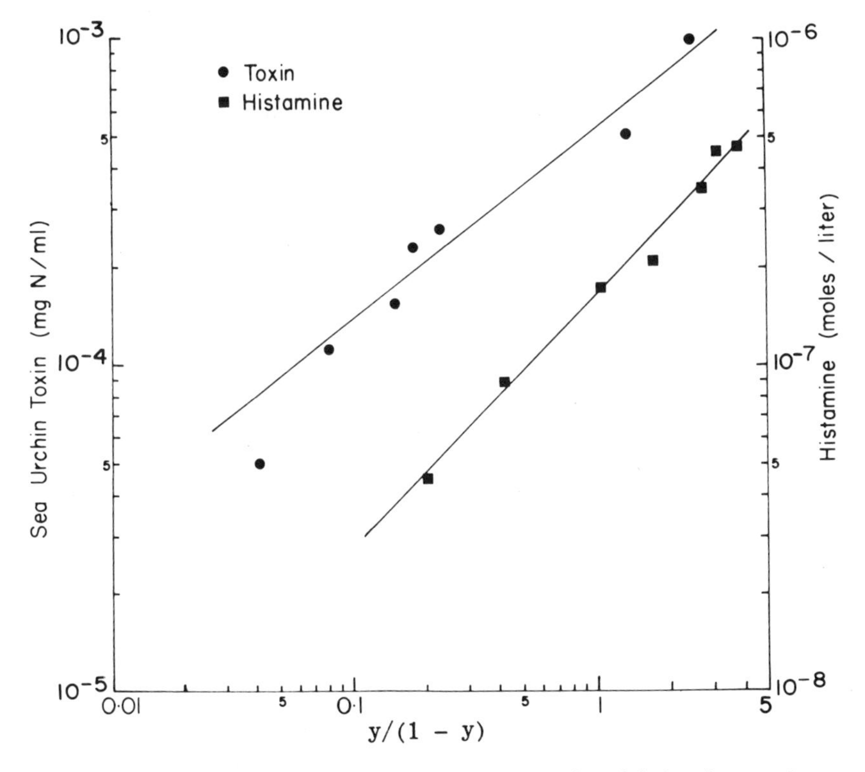

FIG. 3. Logistic dose-response curve for histamine and sea urchin toxin [SUT(61)-C-62]. The median effective dose can be obtained at $y/(1 - y) = 1$ for histamine on the right ordinate and for sea urchin toxin on the left ordinate. (For details see text.)

The dose-response curves were sigmoidal, and the potency of the unknown could be found by interpolation from the standard curve. For more critical work, particularly that which involved the comparison of inhibitors, the sigmoidal functions were linearized by the logistic transformation of von Krogh (Trapani and Feigen, 1959). The typical relationship is exhibited in Fig. 3 for crude toxin and histamine. Since this relationship has the form

$$X = K[y/(1-y)]^{1/n}$$

in which X is the dose of agonist, y is the percent of the maximal response, and 1/n is the slope, the value of K, the median effective dose, is obtained from the value of X at $y/(1 - y) = 1$.

B. Direct Effects of Toxin on the Guinea Pig Ileum

The response of the isolated guinea pig ileum to the direct action of crude toxin preparations is highly variable, and repeated challenge with effective doses of the crude material tends not only to diminish the contractile response to the preparation itself but also to reduce the sensitivity of the tissue to pharmacological agents used to standardize the reaction. In general, the response has a somewhat prolonged induction period (4 to 7 sec), and the contraction develops at a rate dependent on the dose used. Occasionally, moderate doses produce a rapid contraction of the longitudinal musculature; often they initiate a great deal of oscillatory activity either during the rising phase of the tonic contraction or after it has attained its equilibrium amplitude. The gut tends to relax slowly and attains its initial tension only after several washings with toxin-free Tyrode's solution.

1. Blocking Agents

Preliminary experiments with selected blocking agents suggested that pyribenzamine (PBZ) and D-bromolysergic acid (BOL-148) tested in concentrations that completely blocked equipotent doses of histamine and serotonin, respectively, produced only a partial blockade of sea urchin toxin, whereas atropine was virtually ineffective in this respect. Since these results suggested that, in addition to histamine, a variety of other substances such as serotonin, bradykinin, and slow-reacting substance (SRS) might be implicated in the overall response to sea urchin toxin, a more detailed quantitative study was made with a wider range of blocking agents.

Table V presents the mean changes in response and summarizes the degree of block achieved when virtually maximal doses of the several blocking agents were tested against standard doses of toxin. Atropine was entirely ineffective at a concentration double that required to produce a full block to acetylcholine. Pyribenzamine blocked only 20% of the response when it was given in a dose producing complete blockade of the median effective dose of histamine. Mellaril and BOL-148 produced an equivalent block, although the former was matched against an 80% higher dose of toxin than the latter. The most powerful agent was phenylbutazone (ΦBZ), which produced a 97% block against an 86% effective dose of toxin. The dose of toxin in that instance amounted to almost nine times the concentration used in the experiments with atropine, Pyribenzamine, and D-bromolysergic acid.

TABLE V

Effects of Blocking Agents on Response of Guinea Pig Ileum to Sea Urchin Toxin[a]

Sea urchin toxin (mg N/ml x 10^{-5})	Antagonist	Dose (μg/ml)	Response before antagonist (% max)[b]	% Block
2.02	PBZ[c]	0.10	38.5 ± 2.0	20
2.02	Atropine	1.66	38.5 ± 2.0	0
2.05	BOL-148[d]	10.00	32.0 ± 2.0	35
3.60	Mellaril	0.24	44.5 ± 2.0	41
18.60	ΦBZ[e]	1000	86.2 ± 2.0	96.5

[a] SUT (61)-C-62 .
[b] Means and standard deviations based on six experiments.
[c] PBZ is Pyrabenzamine.
[d] BOL-148 is D-bromolysergic acid.
[e] ΦBZ is phenylbutazone.

From these results it is evident that more than one substance is involved in the overall reaction. However, owing to the low discriminatory capacity of the set of blocking agents used, only two legitimate conclusions can be drawn from these data: it is probable that histamine is one of the substances contributing to the response, but acetylcholine is not.

2. Effect of Heat-Degraded Toxin

To determine whether the observed effect was attributable to active toxin, various concentrations equivalent to 3, 7, and 10 LD_{50} doses were heated at 40°C for 20 min before being tested on the guinea pig ileum. The reduction in the intensity of the toxic effect is clearly displayed in Fig. 4.

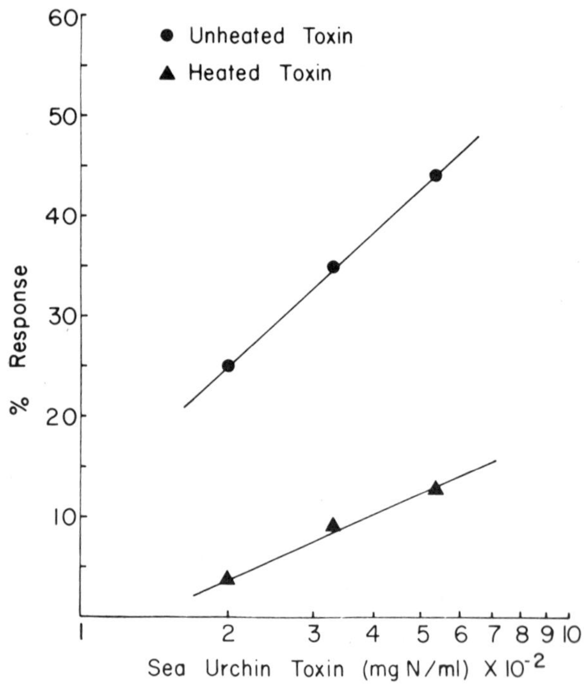

FIG. 4. Effect of heating on potency of sea urchin toxin. Dose-response curve of guinea pig ileum shows loss of activity owing to heating various concentrations of crude SUT at 40°C for 20 min.

C. Evidence for the Release of Physiologically Active Materials from Isolated Tissues

The possibility that the direct action of SUT was due, in part, to the liberation of active materials such as histamine was tested by determining whether the incubation of isolated tissues with SUT could produce dialyzable materials having the same pharmacological effects on test tissues as those described earlier.

Experiments on the release of active materials were made by incubating portions of guinea pig ileum and rat colon with various concentrations of sea urchin toxin at 37°C for given intervals of time as required by the experiment and then dialyzing 2- to 10-ml aliquots of the centrifuged reaction mixture against equal volumes of Tyrode's solution for 48 hr at 4°C. Experiments on the guinea pig heart were made by perfusing standard concentrations of toxin through the coronary circulation and dialyzing selected samples of the effluent.

TABLE VI

Effect of Sea Urchin Toxin[a] on Coronary Flow and Histamine Release in the Guinea Pig Heart

Sea urchin toxin (mg N/ml)	No. hearts	Wet heart weight	Flow (ml/g wet heart/min)	Histamine release[c] (moles/liter/gm wet tissue)
Control	23	1.85 ± 0.55	3.41 ± 0.84	0
2.68×10^{-5} [b]	4	2.31 ± 0.31	0.38 ± 0.10	5.02×10^{-7} ± 0.10
4.85×10^{-5}	4	1.84 ± 0.14	0.29 ± 0.08	1.13×10^{-6} ± 0.23
5.36×10^{-5}	3	1.42 ± 0.12	0.68 ± 0.05	1.40×10^{-6} ± 0.16
6.76×10^{-5}	3	1.21 ± 0.10	0.55 ± 0.11	2.11×10^{-6} ± 0.18
1.08×10^{-4}	3	2.31 ± 0.02	0.81 ± 0.04	2.50×10^{-6} ± 0.10
2.68×10^{-4}	2	2.40 ± 0.10	0.35 ± 0.02	3.00×10^{-6} ± 0.05
3.38×10^{-4}	4	1.46 ± 0.20	0.11 ± 0.09	2.91×10^{-6} ± 0.25

[a] Preparation SUT(61)-C-62.
[b] Values are given as means ± standard deviation.
[c] Histamine estimated chemically.

1. Nature of Materials Released

Dialyzable materials obtained from the attack of active toxin on guinea pig and rat tissues were tested on the isolated guinea pig ileum in the presence of Pyribenzamine, D-bromolysergic acid, and atropine in separate experiments. The data summarized in Table VI suggest that acetylcholine was absent (or destroyed rapidly, if liberated) but that histamine and possibly serotonin -- or some other substance sensitive to BOL-148 -- were definitely present. The presence of histamine in the samples was confirmed by the spectrofluorometric method modified from Shore et al. (1959)

D. Conditions Affecting the Release of Histamine by SUT

The factors determining the release of histamine from various tissues appear to be the concentration of active toxin, the temperature of incubation, and the particular fraction studied.

The dose-dependent release of histamine from the perfused guinea pig heart is evident from the results given in Table VI, and the comparative potencies of materials obtained by ammonium sulfate fractionation are exhibited as dose-response curves in Fig. 5, which shows that most of the histamine-releasing potency of the preparation is found in the material precipitating in the presence of two-thirds saturated ammonium sulfate. The effect of temperature on the release of histamine from the guinea pig lung is evident from the data given in Table VII, which shows the release to be doubled in the range of temperature from 10° to 37°C.

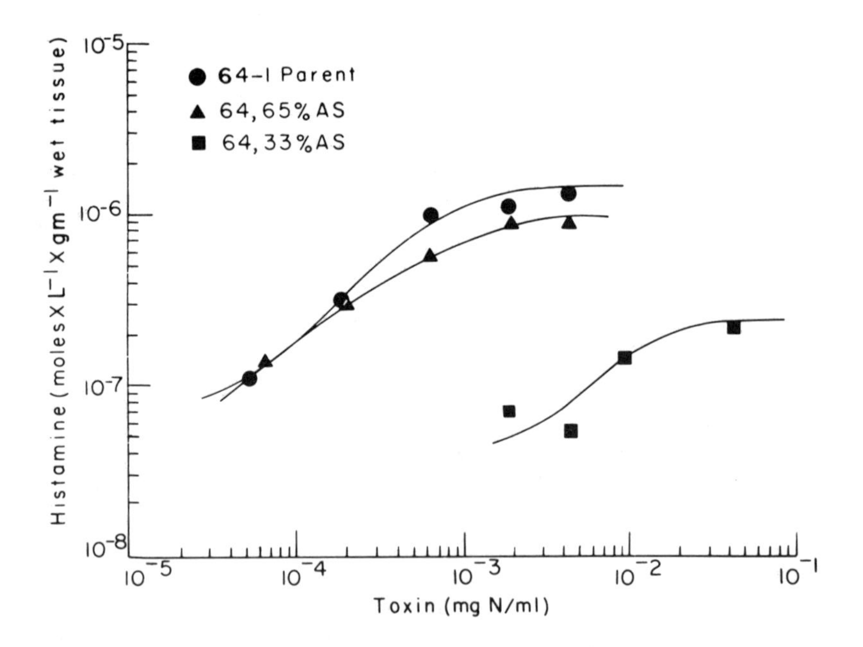

FIG. 5. Comparative histamine-releasing potencies of parent SUT(64), and the two daughter preparations precipitated in the presence of 33% and 65% saturated ammonium sulfate.

TABLE VII

Effect of Temperature on Histamine Released from Guinea Pig Lung by Sea Urchin Toxin[a]

Temperature (^{o}C)	Histamine release[b] (μg/mg/wet tissue) (x 10^{-5})
10	3.29
21	3.51
30	4.67
37	6.87

[a] Incubated with sea urchin toxin [SUT(64)] (4.90 x 10^{-3} mg N) for 20 min.

[b] Chemical estimation.

E. Summary of Pharmacological Evidence

Although the release of histamine, at least, can be proved by the pharmacological and chemical studies, the evidence was very strong that the active fraction had additional properties. In the first place, the crude nondialyzable materials, as well as the active fractions, killed mice. Although mice contain releasable histamine in their tissues, they are generally conceded to be refractory both to endogenously liberated histamine as well as to relatively large doses of the ''authentic'' compound. Second, studies with blocking agents showed that only a part of the activity of the toxin or of the dialyzable materials liberated from tissues could be blocked by Pyribenzamine. Third, the character of the contractile response to the liberated materials, before or after blockade with Pyribenzamine, showed a high degree of delay reminiscent of the action of both SRS and bradykinin. Last, the greater lethality of the crude starting material compared with that of the fractions suggests that a component, for example, an hemolysin or a neurotoxin, which would contribute to lethality but not to the tissue response, might have been removed or destroyed by the fractionation maneuvers.

IV. BIOCHEMICAL STUDIES ON THE ENZYMATIC FORMATION OF PLASMA KININS

During the course of the preceding pharmacological experiments it was observed that the LD_{50} in mice was somewhat lower than the MED required for the direct stimulation of the guinea pig ileum. A plausible reason for this amplification in toxicity is that certain constituents of the active fractions of toxin might have a proteolytic action, which could result in the formation of active plasma kinins of the sort described for the action of snake venom and for the catalysis of serum proteins by kallikrein (Werle and Berek, 1950).

Studies in this section fall naturally into three general domains: (A) the proof that the reaction between SUT with plasma is enzymatic, leading to the identification of the natural substrate; (B) conditions affecting the formation and identification of bradykinin in the reaction product; and (C) the substrate specificity and kinetics of the purified enzyme.

A. Search for the Natural Substrate in Plasma

In the present study we show that the interaction of crude sea urchin toxin, or its purified fractions, with certain substrates in the plasma produces dialyzable, heat-stable substances that have the capacity to initiate contractile reactions of the guinea pig ileum and the rat uterus. A preliminary analysis of the interaction suggests that the toxin behaves kinetically as an enzyme and that its proteolytic action is selective on certain components of the globulin fraction of mammalian plasma.

B. Materials and Methods

1. Bovine Serum Proteins

Whole bovine serum was obtained from the Bakte Scientific Laboratories, Berkeley, California. Before use, 100 ml of serum was dialyzed against 18 liters of 1% NaCl for 48 hr at 4°C. Bovine gamma globulin and bovine serum albumin were obtained from the Armour Pharmaceutical Company. Each preparation was dissolved in 25 ml of 1% NaCl solution. The former was repeatedly precipitated in the presence of one-third-saturated ammonium sulfate (SAS) and the latter in two-thirds-saturated salt at pH 7.8. After the third precipitation each preparation was taken up in 1% NaCl and dialyzed against 30 liters of NaCl solution for 72 hr at 4°C. The dialyzed proteins were centrifuged at 4°C and the supernatants stored at -20°C.

2. Human Serum Proteins

Two kinds of human serum protein fractions were used in these experiments. The preliminary work was done on preparations obtained by salt fractionation, whereas the latter work was performed on commercially available Cohn fractions, some of which were furnished to us as unknowns by the Hyland Laboratories.

The globulins obtained by salt precipitation were prepared according to a method modified from Kendall (1937). Fractions 1 and 2 were obtained in the presence of 1/3 SAS, and Fractions 3 and 4 were obtained in the presence of 1/2 SAS. At each precipitation subsequent treatment produced one water-soluble and one water-insoluble subfraction. Thus Fractions 1 and 3 were water soluble, while Fractions 2 and 4 were water insoluble.

The Cohn fractions of human serum supplied by the Hyland Laboratories as knowns varied in their content of alpha- and beta globulins. According to Pennell (1960), Fraction III-0 was expected to contain 5% alpha- and 84% beta globulin, whereas Fraction IV should have had 89% alpha- and 10% of beta globulins. Subsequently, they supplied Fractions III_1, IV_4, IV_{5+6}, II, and III as unknowns. Electrophoretically pure α_2-M globulin was obtained from Dr. H.H. Fudenberg, University of California Medical Center, San Francisco.

3. Stock Solutions

All the toxin and substrate preparations required for enzymological studies were kept as frozen stock solutions in quantities sufficient for a given day's work so that the activity of the materials would not be affected by repeated thawing and freezing. Stock solutions of sea urchin toxin were made by dissolving 60 mg of the lyophilized powder in 30 ml of 0.15 M sodium phosphate buffer at pH 7. These batches were dialyzed against 6 liters of 1% NaCl for 24 hr and stored in 5-ml bottles at -20°C. The substrate stocks were first extensively dialyzed against 1% NaCl, analyzed for TCA-precipitable N, adjusted to a protein concentration of 5 mg/ml, distributed into 5-ml serum bottles, and stored at -20°C.

4. Enzymological Methods

The stock solutions of toxin were thawed and then suitably diluted with Tyrode's solution immediately before the experiment. The final reaction volumes were 5 ml, and all experiments were made in duplicate. All the reaction mixtures, as well as toxin-free and substrate-free controls, were maintained in a thermostatically controlled Dubnoff metabolic water bath for the time and temperature required by the experimental protocol. At the end of the prescribed incubation period the samples were ''inactivated'' by being placed in a 56°C water bath for 30 min. After inactivation, a 2.5-ml aliquot of each sample was transferred to a cellophane bag. The bag was placed in a 25-ml container and 2.5 ml of Tyrode's solution was added to the vessel. The vials were placed on a rocking device and the active product permitted to dialyze for 24 hr at 4°C. The dialysate, as well as the remainder of the original sample, was frozen at −20°C before bioassay.

5. Pharmacological Estimations

Bioassays were made on the isolated guinea pig ileum, as described previously, or on the sensitized rat uterus, which was primed with stilbestrol.

C. Results

1. Formation and Destruction of Active Materials

Preliminary studies showed that a dialyzable product, having the properties of a plasma kinin, could be produced by incubating whole serum, or certain globulin fractions, with various preparations of the pedicellarial toxin (see Table VIII). No measurable material was elaborated from serum albumin or gamma globulin under these conditions. The toxin preparations were also found capable of inactivating the product formed.

The possibility that the crude toxin might have natural kininolytic activity was tested by incubating samples of synthetic bradykinin with crude toxin. Table IX shows that bradykinin, in quantities sufficient to produce a 50 to 60% response in the isolated ileum, was completely inactivated in 30 to 180 min by the concentrations of toxin used in the experiments reported in Table VIII.

TABLE VIII

Effect of Various Substrates on the Formation of Kinin in the Presence of SUT(64) 65% SAS at 37°C

Preparation	Substrate conc. (mg protein/ml)	SUT conc. (mg N/ml)	Histamine equivalent (moles/liter) (x 10^{-7})
A. Bovine Serum Proteins			
Whole bovine serum	0.4900	0.0863	2.56
Bovine gamma globulin	0.3082	0.0863	0.0
Bovine serum albumin	0.3082	0.0863	0.0
B. Human Serum Proteins			
SAS/3 Kendall Fraction 1	0.3082	0.0863	1.72
SAS/3 Kendall Fraction 2	0.3082	0.0863	4..60
SAS/2 Kendall Fraction 3	0.3082	0.0863	4.34
SAS/2 Kendall Fraction 4	0.3082	0.0863	14.20

TABLE IX

Destruction of Bradykinin by Sea Urchin Toxin 64 at pH 7.5

Test system	Incubation time (min)	Dose tested [a] (ml)	% Reaction	% Inactivated
SUT	30	0.10	0.00	
BK	30	0.10	52.86	
SUT + BK	30	0.10	0.00	100
		0.20	0.00	
SUT	180	0.30	0.00	
		0.40	0.00	
BK	180	0.30	62.22	
		0.40	93.33	
SUT + BK	180	0.30	0.00	100
		0.40	0.00	

[a] The stock concentration of SUT(64) was 0.187 mg N/ml and the stock concentration of synthetic bradykinin was 1 μg/ml.

2. Enzymatic Nature of Product Formation

The results of the preliminary studies gave plausible grounds for inferring that the production of physiologically active materials resulted from an enzymatic attack of one or more substances in the toxin upon a substrate in the plasma. Confirmatory evidence was sought by determining whether such variables as toxin concentration, substrate concentration, and temperature would influence the net kinetics of the system in accordance with this hypothesis.

The following studies in which toxin concentration, serum protein concentration, and temperature are varied show that the toxin behaves kinetically as an enzyme.

D. Toxin Concentration

The effect of toxin concentration on the amount of product formed in 15 min at 37°C was studied by incubating 0.308-mg/ml aliquots of pseudoglobulin, Fraction 4, with five concentrations of the 2/3 SAS fraction of SUT, which ranged from 0.029 to 0.098 mg N/ml at pH 7.5. The highest yield obtained was 10.5×10^{-7} moles/liter (as histamine equivalent). Assuming that as maximal for the array, the results were linearized by the logistic transform and are presented in Fig. 6.

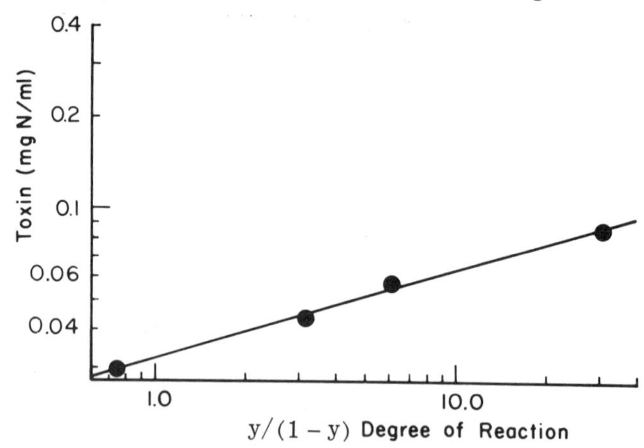

FIG. 6. Effect of toxin concentration on the formation of dialyzable product. The reaction system consisted of 0.308 mg/ml of pseudoglobulin (Fraction 4) and various concentrations of [SUT(64-2)-65%SAS]. The ratio of the percent of total reaction, y, to the remainder, 1-y, is plotted as a function of toxin concentration. The dose of toxin necessary to produce a 50% response occurs at the coordinate when $y/(1-y) = 1$.

E. Substrate Concentration

The relationship between substrate concentration and reaction velocity was studied by determining the time-courses of product formation at 37°C for each of 5 substrate concentrations, ranging from 0.155 to 0.308 mg per ml of pseudoglobulin, Fraction 4, in the presence of 0.093 mg N per ml of toxin. The concentration of product formed at the various time intervals for the entire experimental array gives a family of sigmoidal curves characterized by progressively increasing maxima and diminishing lag periods in response to increasing substrate concentration.

The initial velocity constants for these curves were evaluated from the modified form of the first-order law as applied to growth data by Lotka (1956) and Brody (1945) and to sensitization kinetics by Feigen and Nielsen (1966)... In the present case the working expression can be written

$$1 - P/P_{max} = e^{-k(t - t^*)}$$

in which P_{max} is the maximum product concentration for a given category of substrate concentration, P is the product concentration formed at time t, and t* is the time at which the extrapolated curve crosses the time axis.

Plots of $1 - P/P_{max}$ against $t - t^*$ on semilogarithmic coordinates were linear for each of the time courses, as may be seen in Fig. 7.

The first-order velocity constants were obtained from the slopes of the lines given in Fig. 7 and tested by direct and reciprocal plots for conformance to the Michaelis-Menten relationship. The direct plot showed an upward concavity, indicating that the reaction might have been complicated by a modifier. Accordingly, the data were tested for linearity by determining the goodness of fit to expressions of a higher order. Figure 8 shows that the data are adequately linearized by a plot of the first-order velocity constants against the square of the substrate concentration.

F. Temperature Optimum

Since the toxin preparation had the capacity not only to form the product but to destroy it as well, the possibility existed that the temperature coefficients and the concentration dependence of the two reactions could be such as to confound the estimation of the temperature optimum if only one toxin-substrate system were studied at various temperatures.

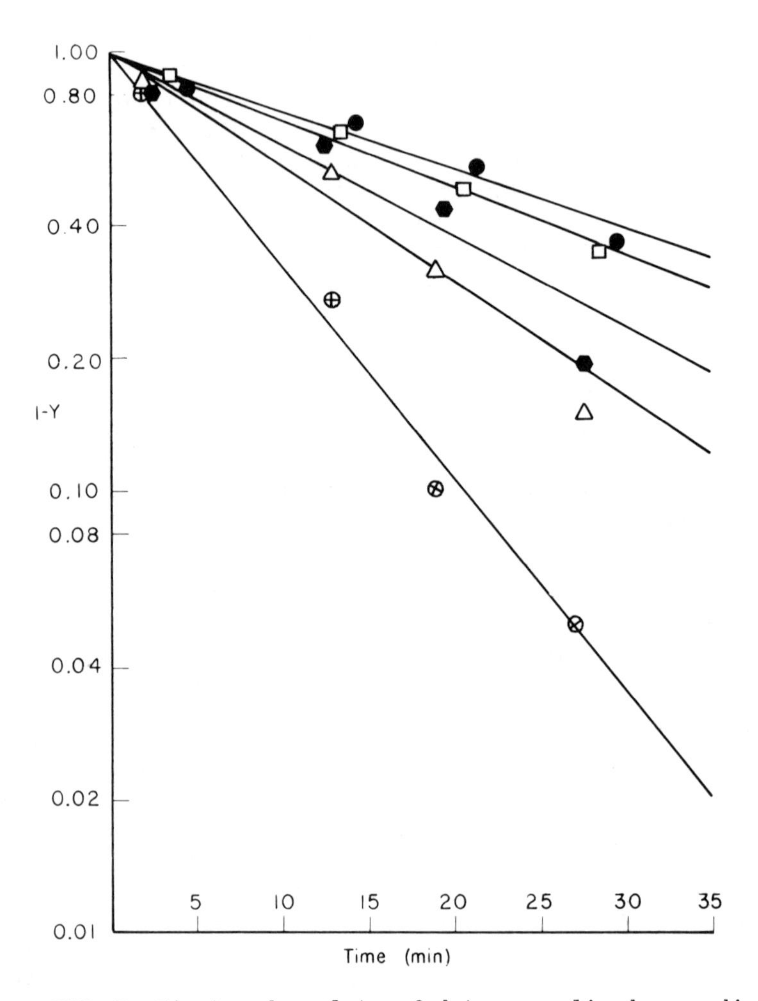

FIG. 7. First-order plots of data normalized according to the procedure described in the text. Concentration of fraction 4: 0.308 mg/ml; 0.217 mg/ml; 0.198 mg/ml; 0.186 mg/ml; and 0.155 mg/ml.

For this reason the temperature optimum was studied by measuring the net production of active material for an array of six toxin concentrations at each of the four temperatures. The substrate was 0.20 mg/ml (Cohn Fraction IV_4) in all cases, but the toxin ranged from 0.04 to 0.40 mg/ml for each of the four temperatures: 19°, 26°, 37°, and 41°C.

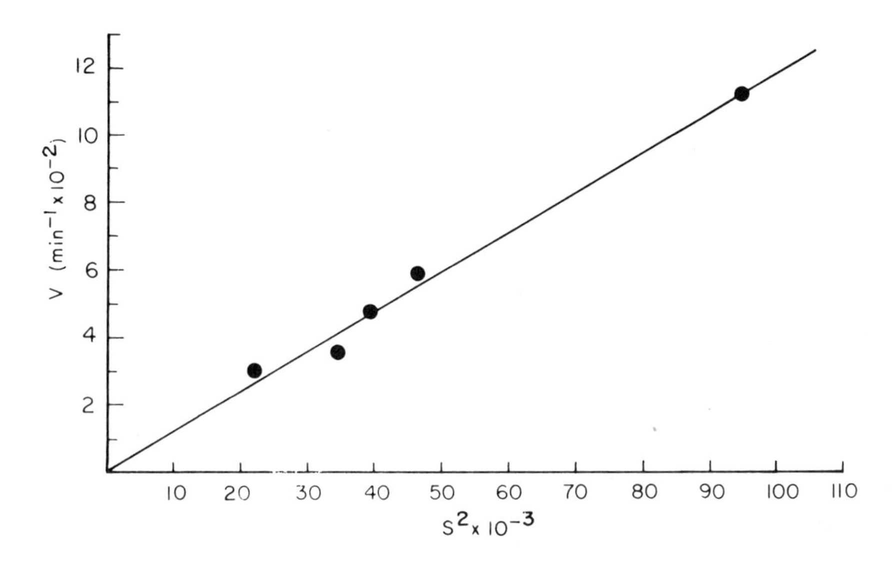

FIG. 8. The variation of the first-order velocity constant (V) obtained from the slopes of the lines in Fig. 7, with the square of substrate concentration (S^2).

The initial velocities of the several reaction systems were obtained in the usual way. The maximum velocity was observed at 26°C. The data exhibited in Fig. 9 are expressed as percentages of maximal velocity at 26°C.

3. Identification of the Natural Substrate

Studies of substrate specificity showed that at least two serum proteins could serve as substrates for the toxin preparations. The beta globulin-rich preparations were most effective in producing materials active on guinea pig tissues, but they yielded no measurable product as judged by assays on rat uterus. On the other hand, alpha-rich materials yielded products effective on tissues of both animals, and it was subsequently shown that electrophoretically pure α_2-M globulin, tested in appropriate concentrations, could quantitatively account for the product yielded from the alpha-rich, but impure, Cohn fractions.

A preliminary assessment of substrate effectiveness was made by comparing the concentrations of product yielded by the attack of toxin on Cohn fractions selected for their differing content of alpha- and beta globulins. Fraction III-0 was ex-

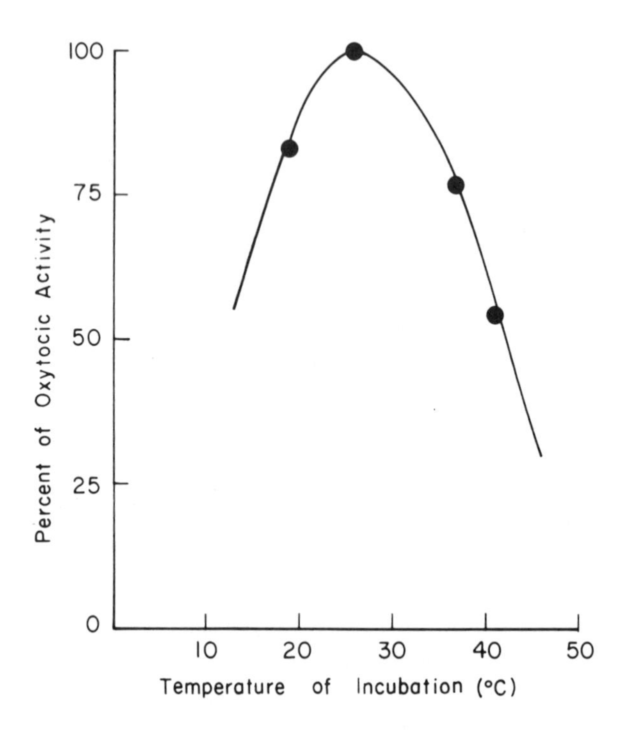

FIG. 9. Temperature optimum for the production of maximum oxytocic activity from the interaction of SUT(65) Sephadex Fraction I and Cohn Fraction IV.

pected to contain about 84% beta globulins (68% β_1 and 16% β_2) and about 5% mixed alpha globulins, whereas Fraction IV-1, according to Pennell (1960), normally contains 89% mixed alpha-globulins and 10% mixed beta globulins.

The results given in Table X showed that more bradykinin-like activity was elaborated by Fraction III-0 than by Fraction IV-1 with respect to the guinea pig ileum, although it was about the same for the rat uterus in both cases. Both substrates were considerably more effective than diluted beef serum, even though the latter was present in approximately a 20-fold greater protein concentration than the Cohn fractions.

It appeared possible that the difference between the two assays might have been a reflection of the attack of toxin on different substrates, one furnishing material effective on guinea pig gut and the other producing substances active only on the rat uterus. The effect of toxin concentration on kinin production by the attack of SUT on III-0 was studied with re-

TABLE X

Comparative Effectiveness of Cohn Fractions and Bovine Serum as Substrates for SUT(64) 65% SAS According to Various Bioassays[a]

		Enzyme	Guinea pig ileum		Rat uterus
Substrate[b]	Substrate conc. (mg protein/ml)	SUT(64) 65% SAS (mg/ml)	Bradykinin equivalent (x 10^{-5} mg/ml)	Histamine equivalent (x 10^{-7} mg/ml)	Bradykinin equivalent (x 10^{-5} mg/ml)
III-0	0.40	0.40	10.50	5.00	1.110
III-0	0.40	0.20	7.30	4.00	0.780
IV-1	0.40	0.40	6.35	2.80	1.140
IV-1	0.40	0.20	5.33	2.50	1.000
Beef serum	1:10	0.40	2.83	1.73	0.483
		0.20	0.93	0.58	0.000

[a] Incubation period: 20 min; temperature: 37°C; inactivated: 56°C for 30 min.

[b] Fraction III-0 was high in beta globulins and Fraction IV-1 was high in alpha globulins. For approximate composition see text.

spect to the activities of the product on guinea pig ileum and rat uterus. The results displayed in Fig. 10 show that the dependence of product formation on toxin concentration is much steeper in the guinea pig assay than in the rat uterus test. The observed results suggest that the two major molecular species in Fraction III-0 furnish qualitatively different products when attacked by an ensemble of enzymes in the toxin preparation and that the difference in amounts of material formed is a reflection of the quantities of the various substrates present.

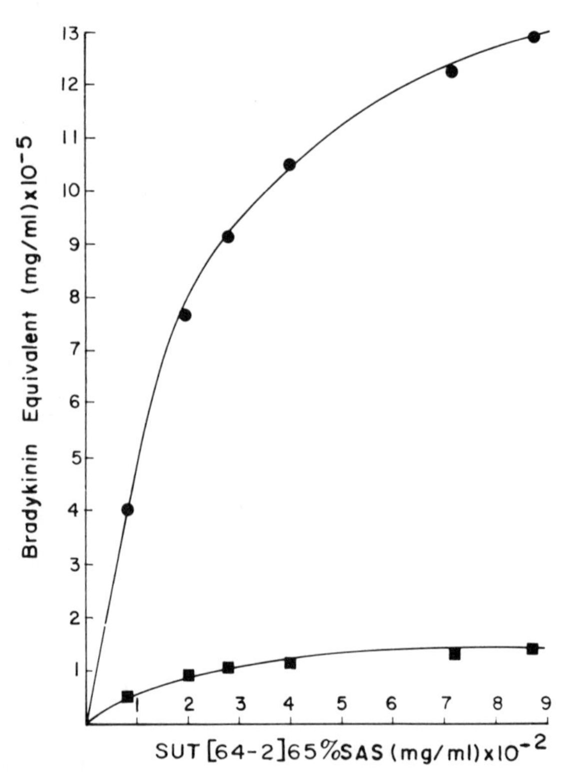

FIG. 10. Comparative sensitivity of guinea pig (●——●) and rat (■——■) tissues to the product formed from Cohn Fraction III-0 and various concentrations of [SUT(64–2)–65%SAS]. The reactions were carried out at 37°C for 20 min and then inactivated at 56°C for 30 min before testing. The material was assayed on guinea pig ileum in Tyrode's at 37°C and on the rat's uterus in de Jalon's solution at 30°C.

Pilot tests indicated a considerable gain in the rat uterus assay whenever the concentration of materials having alpha globulins was raised. The validity of these observations was tested by experiments on unknowns prepared through the courtesy of Mr. Kingdon Lou of the Hyland Laboratories, who furnished materials differing in their content of alpha- , beta-, and gamma globulins. The reactions were carried out for 10 min at two substrate and three toxin concentrations at 30°C. The results were submitted to Mr. Lou, who then identified the samples as being II + III(β,γ); $IV_4(\alpha,\beta)$; III(α,β,γ); and $IV_{5+6}(\alpha)$.

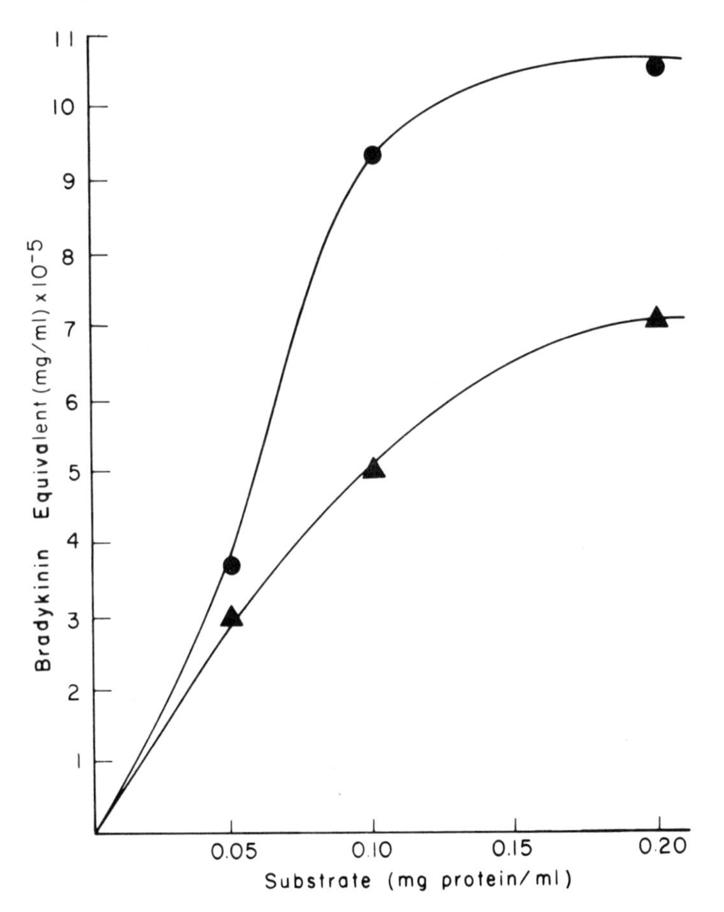

FIG. 11. Relative potencies of Cohn Fractions IV_4 (▲—▲) and IV_5 and IV_6 (●—●) as substrates for SUT(65) Sephadex Fraction I (9.2×10^{-3} mg N/ml).

TABLE XI

Effect of Various Cohn's Fractions as Substrates in the Formation of Kinin in the Presence of Sea Urchin Toxin SUT(64) 65% SAS at 30°C [a]

Sample code (unknown)	Cohn fraction (disclosed)	Enzyme conc. SUT(64) 65% SAS (mg/ml)	Substrate conc. (mg protein/ml)	Incubation time (min)	Bradykinin equivalent	
					GP ileum (mg/ml)	Rat uterus (mg/ml)
1	II, III	0.080	0.200	10	0.00	0.00
2	IV_4	0.080	0.200	10	5.67×10^{-5}	6.05×10^{-6}
3	III_1	0.080	0.200	10	1.00×10^{-5}	0.00
4	IV_5, IV_6	0.080	0.200	10	6.62×10^{-5}	5.30×10^{-6}
1	II, III	0.160	0.200	10	1.00×10^{-5}	0.00
2	IV_4	0.160	0.200	10	9.00×10^{-5}	7.30×10^{-6}
3	III_1	0.160	0.200	10	1.87×10^{-5}	0.00
4	IV_5, IV_6	0.160	0.200	10	7.25×10^{-5}	8.35×10^{-6}
1	II, III	0.160	0.400	10	2.13×10^{-5}	0.00
2	IV_4	0,160	0.400	10	10.80×10^{-5}	1.78×10^{-5}
3	III_1	0.160	0.400	10	2.02×10^{-5}	0.00
4	IV_5, IV_6	0.160	0.400	10	12.50×10^{-5}	1.50×10^{-5}
1	II, III	0.280	0.400	10	2.09×10^{-5}	0.00
2	IV_4	0.280	0.400	10	9.52×10^{-5}	1.07×10^{-5}
3	III_1	0.280	0.400	10	1.98×10^{-5}	0.00
4	IV_5, IV_6	0.280	0.400	10	10.00×10^{-5}	8.5×10^{-6}

[a] Temperature of incubation: 37°C; incubation time: 10 min; inactivation temperature: 56°C for 30 min; globulin composition of samples: 1, β + γ; 2, α + β; 3, α, β, γ; 4, α.

Inspection of Table XI shows that the beta-rich substrates furnished no measurable product in tests on the rat uterus, whereas the alpha-rich materials yielded active substances that varied in a direct relation to the substrate concentration used in the reaction, except at the highest toxin and substrate concentrations in which a certain amount of degradation took place.

In order to obtain a quantitative confirmation of the difference between the alpha and beta globulins, Fraction IV_4 and IV_{5+6} were tested at lower substrate concentrations in the expectation that by reducing the reaction velocity the differences between the two substrates would become more critical. Since the alpha globulin content of Fraction IV_4 is 40%, whereas that of Fraction IV_{5+6} is 85%, the data exhibited in Fig. 11 evidently support the hypothesis concerning the specificity of the alpha globulins as substrates in the reaction.

Conclusive proof of this argument would require positive tests with electrophoretically purified α_1- or α_2-globulins and negative tests with comparable preparations of β_1- and β_2-globulins. Unfortunately, the only preparation available in pure form at that time was α_2-globulin. The results of comparative tests in which immunoelectrophoretically pure α_2-globulin and Fraction IV_{5+6} were studied at 4% concentrations are shown graphically in Fig. 12. It is evident from these results that the α_2-M molecule is a substrate in the reaction with SUT, which can form products having an effect on the rat uterus.

G. Relative Enzymatic Potencies of Sephadex Fractions

A test of the comparative enzymatic potencies was made on the three Sephadex preparations and the 2/3-SAS fraction by measuring their ability to form active material during the reaction with α_2-M globulin. Samples of α_2-globulin, made up to 0.10 mg protein/ml, were incubated for 10 min with 1.87 x 10^{-2} mg N/ml aliquots of the various toxin fractions mentioned earlier. The reactions were stopped by heating the samples at 56^oC and the products bioassayed on the guinea pig ileum.

The results showed that the greatest yield of active products, estimated as bradykinin, was given by the fraction precipitated in 2/3 SAS. So far as the Sephadex preparations were concerned, the greatest net yield was given by Fraction III, 1.4 x 10^{-5} mg/ml (as bradykinin), and the least (4 x 10^{-6} mg/ml) by Fraction I. In view of the fact that the lethal activity of the Sephadex preparations was concentrated in Fraction II, it appears evident that there is no simple relationship between lethality and enzymatic potency with respect to this substrate.

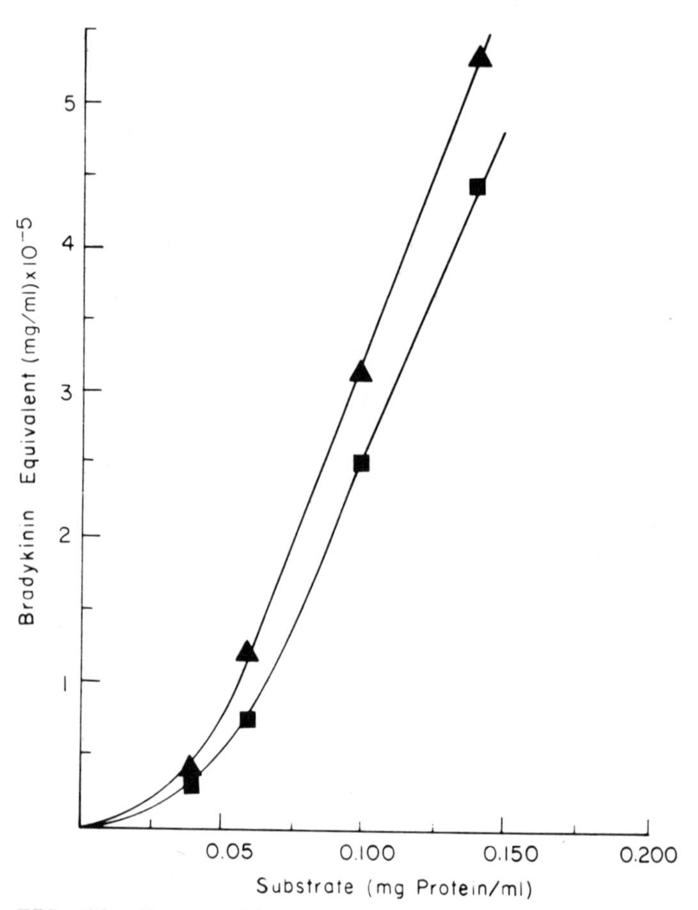

FIG. 12. Comparative potencies of Cohn Fraction IV_5 and IV_6 (▲——▲) and pure α_2-M globulin (■——■) as substrates for 1.87×10^{-2} mg N/ml of [SUT(64-2)-65%SAS].

H. Comment

The present studies have produced convincing evidence that the reaction between sea urchin toxin and plasma proteins results from an enzymatic attack of toxin molecules on specific substrates to yield dialyzable products that have the capacity to stimulate smooth muscle.

The active constituents of sea urchin toxin are proteins, which react with α_2- or β-globulins to form materials active on the rat uterus or guinea pig ileum, respectively. In addition to these proteolytic effects, the toxin preparations were also capable of destroying the product and of inactivating synthetic bradykinin. To add to the complexity of the system, the plasma protein fractions used as substrates were shown to form active materials and to inactivate the products in the <u>absence</u> of toxin.

The simultaneously competitive reactions obviously confounded any attempt to demonstrate a clear kinetic order in the present system, and it was clear that the problems here were the usual ones found in the initial experiments with the kallikrein system in which both the kininogenic and kininolytic effects, exhibited by both the crude ''enzymes'' as well as the crude ''substrates,'' obscured the relationship between the reactants and the products formed.

Taking first the case of the enzyme, we should expect the dependence of reaction velocity on enzyme concentration to be linear. The dependence observed experimentally, although linear at low concentrations of toxin, appeared to decelerate as the toxin concentration was increased, presumably because of kininolytic behavior of the toxin.

Considering next the influence of substrate concentration on the reaction velocity, we see that the dependence of the first-order velocity constants on the concentration of pseudoglobulin does not give the usual hyperbolic relationship. The present data could be linearized only when the first-order velocity constants were plotted against the square of substrate concentration. According to Frieden (1964) such a complexity can arise if the substrate contains ''modifiers,'' i.e., if the activity of the enzyme is markedly affected by compounds that are not substrates for the reaction of interest. In the present case it may be assumed that substrate activation could have accounted for the complexity of the function.

The substrate preparation used was capable of forming kinins spontaneously. That finding is in accord with prior reports showing plasma protein fractions contain inactive kallikreins that are easily activated and thus capable of attacking α_2-globulins to form plasma kinins. It is significant that the kinin-forming system can be rapidly generated at neutral pH by procedures as simple as dilution or exposure to wettable surfaces.

The substrate fractions also contain kininases, and although these were presumed to be inactivated with acid in this connection, it is useful to recall that treatment with acid not only inactivates the kininases, but also promotes the activation of an enzyme system that spontaneously forms kinins.

An additional factor, unknown at the time of the experiments, was that more than one substrate in the crude fraction could have been involved in the reaction. It is entirely possible, as shown by Jacobsen (1966), that the plasma kallikrein could have acted on at least two substrates, and it is evident from our studies that both α_2- and β-globulins are capable of being attacked by SUT. Whether or not these substrates are attacked by the same or by a different species of molecule has not yet been established.

The bioassay does not distinguish physiologically between those kinins and the dialyzable products formed from the attack of toxin on the substrate; thus, it is not possible to assess whether, for example, the toxin preparations could have served to activate the plasma kallikreins at the same time that they were attacking α_2-globulins.

Finally, the toxin can degrade synthetic bradykinin. Since bradykinin is one of the principal reaction products of the plasma kallikrein system, it is probable that the net output of active material from the present reaction system must have reflected a certain degree of bradykinin destruction.

I. Identification of Bradykinin in the Reaction Product

To resolve some of the questions raised by the foregoing evaluation it was necessary to show that bradykinin was, in fact, formed during the reaction. If the compound was produced, it would then be legitimate to inquire which fraction(s) produced it by studying their kinetics, substrate specificity, and point of attack.

The identification of bradykinin in the peptide residues remaining after the digestion of heated plasma with trypsin has been described by Prado(1964). In the present case the plan of attack was to compare the pharmacological activity of the peptides recovered from digestion of the substrate with SUT with those obtained by trypsin digestion. If the biological tests suggested that bradykinin was formed in the reaction, the peptides would then be purified further by filtration through Sephadex and, finally, identified by the comparative chromatography of known bradykinin and the unknown peptide on paper.

J. Conditions Optimizing the Production of Active Materials

Owing to the presence of powerful kininases in the cruder toxin and substrate preparations, our aim was to maximize the formation of the product by selecting the suitable time and temperature for the reaction. These conditions were determined by studying the time course of the net output of material as a function of toxin concentration at various temperatures.

1. Toxin Concentration

Human plasma was heated to 58°C for 3 hr and dialyzed, after centrifugation, for 18 hr against 1% NaCl in the cold room. Four-milliliter aliquots of this substrate (HTP) were treated with 1-ml portions of crude toxin (SUT-67) so that the final concentrations of toxin ranged from 6.05×10^{-3} to 7.04×10^{-1} mg N/ml. The test samples, as well as the appropriate controls, were incubated for 10 min, heated to 56°C, and dialyzed overnight against 4 ml of 1% NaCl in closed vessels, which were mounted on a shaker in the cold room. It is evident from Table XII that the optimal reaction temperature is 32°C for a 16 min incubation for the production of bradykinin-equivalent material (BKE). Under these conditions, even the lowest toxin concentration appeared to be highly effective in tests on the rat uterus.

TABLE XII

The Effect of SUT-67 Concentration and Temperature on the Formation of Dialyzable Kinin [a]

SUT concentration (mg N/ml)	Bradykinin equivalents formed [b]		
	27°C	32°C	37°C
6.5×10^{-3}	9.1	14.0	3.1
1.3×10^{-2}	–	16.0	8.0
2.6×10^{-2}	12.5	14.0	13.0
5.2×10^{-2}	17.0	20.0	11.0
1.04×10^{-1}	20.0	20.0	13.0

[a] Digestion time 10 min.
[b] μg/ml $\times 10^{-2}$.

2. Temperature

The time course of BKE formation during the digestion of HTP in the presence of 0.026 mg N/ml SUT-67 was studied at 27°

32°, and 37°C. The amount of BKE formed increased monotonically with time at the lower temperatures, but at 37°C the amount of BKE formed reached a maximum in 10 min and then declined rapidly.

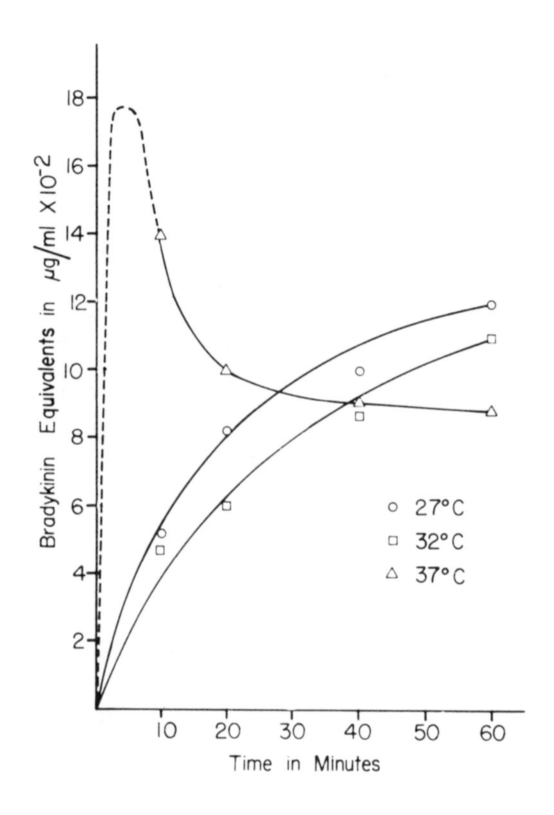

FIG. 13. Time course of BKE formation from heat-treated plasma at 27°, 32°, and 37°C as assayed on rat uterus.

3. Kininolysis

Evidence that the increased kininase activity could reduce the net output of active material was sought by direct tests on the inactivation of synthetic bradykinin at 27°, 32°, and 37°C. These results are shown in Table XIII. They confirm that the toxin has a rapid destructive effect on this peptide. The degradation was somewhat greater at 27°C than at the other two temperatures. Since more than one material evidently is produced in the attack of SUT on HTP, these results only partially account for the time courses in the preceding experiment.

TABLE XIII

Temperature and the Rate of Digestion of Bradykinin by SUT-67 as Assayed on Rat Uterus

BK(4 x 10^{-2} μg/ml) SUT-67(2.6 x 10^{-2} mg/ml)	Temp. (°C)	Time in min			
		10	20	40	60
BKE digested μg/ml x 10^{-2}	27	3.40	3.58	3.81	4.00
	32	3.00	3.53	4.00	–
	37	2.80	3.32	3.68	3.83
Percent BKE digested	27	85.00	89.00	95.00	100
	32	75.00	88.40	100	–
	37	70.00	83.00	92.00	98.3

K. Conditions Optimizing the Recovery of Kinins

Prado et al. (1965) recovered peptides that had formed during the reaction between plasma and either trypsin or ficin by means of a solvent extraction similar to the one described subsequently in this chapter. This procedure is time consuming and laborious for quantitative recovery, particularly if only small quantities of product are sought for biological assay. In addition, the direct solvent method has inherent uncertainties as to the completeness of the recovery, owing to possible selectivity of the solvent system for one or another of the active principles. The active materials were known to be dialyzable, and it therefore appeared useful to prepare a moderately large batch of active starting material by dialysis and to test various methods of purification on the dialyzable products rather than on samples of the digestion mixture itself.

To test the various alternatives in the isolation process it was first necessary to establish the conditions that would maximize the production of kinin and to establish the specific activity of the product, i.e., the kinin equivalent per unit N. Next, the efficiency of solvent recovery had to be compared to that of gel filtration on Sephadex; and, lastly, the active material obtained from the SUT digest had to be separated by chromatography on paper and the biological activity of the various loci tested and identified.

1. Recovery by Dialysis

Three hundred milliliters of a reaction mixture containing 2.5 gm protein N of heated plasma substrate and 15.60 mg toxin N were incubated at 27°C for 60 min. The reaction was stopped by heating the system for 30 min and the kinins formed were dialyzed against three 1500-ml portions of water. The dialysates were concentrated to 110 ml on a rotary evaporator, and samples of the material obtained before and after dialysis were studied for recovery of nitrogen and for pharmacological activity. The data given in Table XIV show the recovery of 37.5% of pharmacological activity against a 0.64% recovery of total N, an obviously effective separation.

TABLE XIV

Recovery of Dialyzable Kinins from the Digestion of HTP

Sample	Vol. (ml)	BKE (μg/ml) (x 10^{-2})	Total BKE (μg)	N (mg/ml)	Total mg N recovered	$\frac{\mu g\ BKE}{mg\ N}$
HTP-SUT-67 digest	300	5.5	16.5	10.34	2,542	6.5 x 10^{-3}
Pooled concentrated 1st and 2nd external dialysate	110	5.6	6.2	0.153	16.83	3.7 x 10^{-1}
% Recovery			37.5		0.64	

2. Comparison of Solvent Extraction with Gel Filtration

The active kinins prepared by dialysis were used for comparative study of the efficiencies of solvent extraction and filtration in the following way. A 25-ml sample was purified by solvent extraction and precipitation according to the technique of Prado, and a separate 5-ml aliquot of the external dialysate was passed upward through a column of Sephadex G-25. In the latter case, the contents of all tubes containing biologically active material were pooled, evaporated to dryness *in vacuo,* and taken up in 1.0 ml of water for study.

The results of the recovery experiments, summarized in Table XV, suggest that although there was no evident advantage of the gel filtration over solvent extraction in the recovery of total nitrogen (16.6% as against 14.6%), the material prepared by gel filtration had approximately twice the physiological potency of the preparation obtained by the Prado method. Since 100% of the biological activity was recovered by gel filtration but only 54% by the Prado extraction for an almost identical (85%) loss of nitrogeneous material in both cases, it is not clear whether the difference in specific potency (μg BKE/mg N) occurred because of inactivation of kinins by the solvents used for extraction or because of differences in the selectivity of the two methods.

TABLE XV

Recovery of Dialyzable Kinins Obtained by Solvent Extraction and Gel Filtration

	Sample volume (ml)	Input			Recovery			
		Total mg N	Total μg BKE	μg BKE / mg N	Total mg N	Total μg BKE	μg BKE / mg N	% BKE recovered
Solvent	25	3.83	1.4	0.36	0.56	0.75	1.33	54
Sephadex G-25	5	0.77	0.28	0.36	0.12	0.28	2.39	100

3. Resolution of Peptides and Identification of Bradykinin by Chromatography on Paper

Five 10-μl aliquots of the Sephadexed unknown were applied to a Whatman No. 1 filter paper along with a bradykinin standard prepared by passing 0.5μg of the synthetic product (Sandoz No. 69055) through Sephadex G-25 under conditions identical to those used for the unknown. The paper was dried and one portion was developed by spraying a solution of 0.2% ninhydrin in acetone. Figure 14 shows the digestion mixture to be resolved into seven components, three of them corresponding to the spots shown by the synthetic product. Appropriate sections were then cut out from the unstained portion of the filter paper, and each spot was extracted in 1 ml of Tyrode's solution for bioassay on the guinea pig ileum. The greatest activity, as shown in Fig. 15, was found in loci 4 and 5, corresponding to the slower moving components in the commercial preparation.

FIG. 14. Paper chromatogram on Whatman No.1 filter paper of authentic bradykinin (BK) and of dialyzable product obtained from the digestion of plasma substrate by SUT-67 (SUT-HTP).

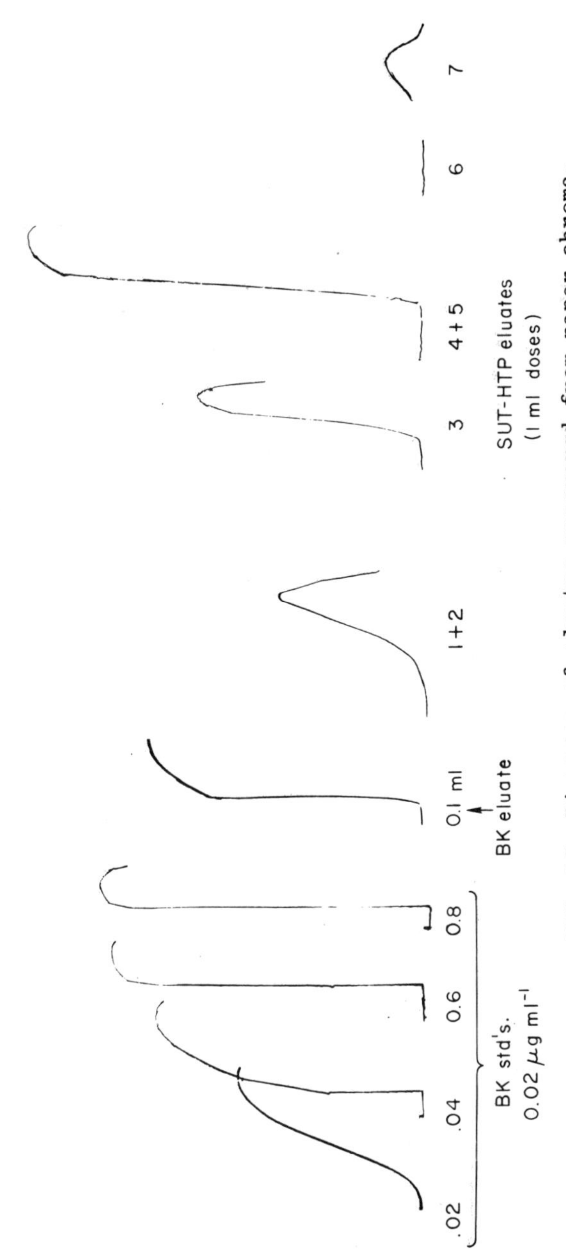

FIG. 15. Bioassay of eluates recovered from paper chromatogram shown in Fig.14.

L. Comparative Efficiency of SUT and Trypsin as Kininogenases

The bioassay results exhibited in Fig. 15 showed that other active peptides were produced in the reaction along with bradykinin. Since none of the pharmacological tests for bradykinin is completely specific, it appeared plausible that the unidentified materials could account for differences in titer often found between assays made on various isolated organ preparations in different species of animals. Similar problems had been encountered by Prado in comparative studies of kinin formation by the attack of ficin and trypsin on heated plasma, and for that reason his operating control was the ratio of potencies between the ileum and the uterus.

To determine whether there were significant differences in the distributions of peptides by SUT and trypsin from the same substrate, we prepared large scale batches of active peptides by the Prado solvent extraction method, using trypsin as the enzyme in one case and SUT in the other, according to the method described in the next paragraph.

Active peptides for pharmacological and chemical studies were produced by treating HTP substrate with preparations of SUT or trypsin (salt-free Trypsin No. 9680, NBC twice crystallized). The enzyme or toxin preparations were made up in 100 ml of glucose-free Tyrode's solution and added to 250 ml of HTP. The reaction proceeded for 10 min at 30°C. At the end of the incubation period the mixture was poured into 1000 ml of boiling ethanol and heated at 80°C for 30 min. The copious protein precipitate was removed by filtration and centrifugation of the filtrate. The supernatant was evaporated to dryness on a rotary evaporator in a water bath at 50°C for 18 hr. The residue was taken up in 25 ml of distilled water and centrifuged. The lipid layer and the pellet were saved for a second extraction and the supernatant was saturated with 8.9 g of NaCl. This solution was brought to pH 1.5 with 4 N HCl, and an equal volume of butanol-water was added. The butanol-water mixture was thoroughly mixed and then separated by centrifugation. These extracts were pooled, added to 400 ml of anhydrous peroxide-free ether, and stored overnight on ice. The precipitate that formed was collected by centrifugation and dried over anhydrous $CaCl_2$ *in vacuo*.

The SUT digest yielded 130.2 mg and the trypsin reaction produced 70.5 mg of ether-precipitable material, respectively. Table XVI gives the comparative yields of kinin formed by the actions of trypsin and of SUT according to the three assays carried out against bradykinin standards. On the basis of dry-

TABLE XVI

Yields of Ether-Precipitable Kininlike Materials Obtained by Digestion of HTP

Assay method	Sea urchin toxin			Trypsin		
	Weight of material recovered (mg)	μg (as BK) by bioassay	μg BK per mg	Weight of material recovered (mg)	μg (as BK) by bioassay	μg BK per mg
Guinea pig ileum		2.47	0.019		3.67	0.052
Rat uterus	130.19	1.30	0.010	70.48	0.54	0.008
Rat duodenum [a]		0.28	0.004		0.28	0.004
Ileum/uterus		1.9			6.8	

[a] Bradykinin relaxes rat duodenum but contracts the other smooth muscle preparations.

weight ratio, the trypsin reaction produced about 2.5 times the guinea pig active material obtained by the SUT method, although the titers by the rat tissues were nearly identical.

The differences in sensitivity found between guinea pig and rat tissues are brought out even more sharply in experiments made on perfused hearts. The results, presented as a summary in Table XVII, show first of all that synthetic bradykinin produces diametrically opposite results in the hearts of the two species with respect to flow and amplitude. In the guinea pig heart the flow rate is increased and the amplitude is decreased, whereas in the rat organ these reactions are reversed. The heart rate is not affected in either case. The action of the trypsin digest is substantially the same as that of bradykinin with the exception of a slight decrease in amplitude noted in the rat heart tested with the natural product. Tests with the SUT digest show, qualitatively, the same direction of changes in coronary flow rate. The principal point of distinction between the effects of bradykinin or trypsin digest, on the one hand, and the action of the SUT products, on the other, lies in the severe reduction of the heart rates of both species of animal. These findings, along with the assay presented in Table XVII, give a clear suggestion that although bradykinin might be produced by both enzymes, SUT had additional points of attack on the natural substrate, which yields a greater diversity of dialyzable materials than those produced by crystalline trypsin.

TABLE XVII

A Comparison of the Effects of Bradykinin, Trypsin-Digest, and SUT-Digest

Tissue		Test substance					
		Bradykinin (% change)		Trypsin digest (% change)		SUT digest (% change)	
Parameter	BK dose (μg)	GP	Rat	GP	Rat	GP	Rat
Coronary flow	2	+30	−20				
	10	+56	−12	+90	−46	+71	−87
Amplitude	2	−25	+12				
	10	−15	+ 9	− 8	−14	−40	−75
Rate	2	0	0				
	10	0	0	0	0	−50	−25

M. Comment

The question naturally arises as to the differences observed between the pharmacological effects of the SUT- and trypsin-HTP digests obtained by solvent extraction. The evident explanation is that, with the exception of bradykinin, the array of other peptides produced in the two cases might have differed qualitatively or quantitatively. In support of this view is the fact that the assay values obtained on the guinea pig ileum tend to be higher than those obtained on rat tissues. The range of sensitivity of guinea pig ileum is much broader than that of the rat uterus. Even though its range of response is limited, the rat uterus contracts to many peptides including bradykinin, kallidin, and eledoisin. The rat duodenum distinguishes between bradykinin and eledoisin by relaxing to the former and contracting to the latter. In the present case, both the SUT and trypsin digests relaxed the duodenum, indicating that bradykinin was one of the components present in the medium. Assays on the isolated heart showed several points of difference between the products formed by trypsin and those formed by SUT and suggested that the latter product may have contained a mixture of peptides, one of which was bradykinin.

The present section will deal with the enzyme specificities and kinetics of several of the components of SUT with relation to their lethal toxicity and hemolytic action.

N. Materials and Methods

1. Toxin Preparations

Three general types of toxin preparation were used in this study: (a) crude pedicellarial extract (SUT-67), (b) material that had been fractionated by gel filtration on Sephadex G-200, and (c) fractions prepared by eluting the absorbed material from hydroxylapatite. These were prepared according to the methods described in the initial portion of this chapter.

2. Substrates and Inhibitors

Heat-treated human plasma was used as the substrate in experiments involving the digestion of plasma proteins by sea urchin toxin. The substrates used for identifying the points of attack of the various toxin preparations were azocasein, p-tosyl-L-arginine methyl ester (TAME), and N-acetyl-L-tyrosine ethyl ester (ATEE) (Mann Research Laboratories, New York City). Synthetic bradykinin (BK) was obtained through the courtesy of the Sandoz Pharmaceuticals, Hanover, New Jersey, as ampoules containing 0.1 mg of peptide/ml. The proteinase inhibitor ''Trasylol'' was supplied as preparation A-128 by Metachem, Inc., New York.

3. Enzymological Procedures

The proteolytic behavior of the purified SUT fractions was assessed by determining their ability to digest heated plasma and azocasein. The substrate in the latter case was 1.3% sulfanilamide-azocasein prepared according to the method of Charney and Tomarelli (1947). Tests were carried out according to the method of Markus and Werkheiser (1964) by incubating 1-ml aliquots of the 1.3% azocasein dissolved in 0.1 μ borate buffer (pH 8.5) with 0.5-ml portions of selected toxin preparations at 32°C. The reaction was stopped after 60 min by precipitating the undigested azoprotein with 1.0 ml of 1.12 M perchloric acid and the degree of digestion was determined by measuring the optical densities of the supernates at 390 nm.

The kinetic behavior of SUT toward two synthetic esters, TAME and ATEE, was also evaluated and compared with the effects of twice-crystallized salt-free trypsin and thrice-crystallized alpha chymotrypsin, both from the Nutritional Biochemical Co. The continuous titration method employed was modified from the technique described by Markus et al. (1967) for measuring the rate of tryptic digestion of serum albumin. This method measures the rate of substrate digestion by the continuous neutralization of the H+ ions liberated by the enzymatic cleavage of the substrate ester bonds. The reaction was carried out in a 30-ml vessel, jacketed to permit circulation of water from a 30°C bath. The vessel contained the substrate solution and accommodated a pair of pH meter electrodes, as well as the tip of a syringe microburette for the delivery of base.

O. Results

1. Attack on Natural Substrates

The degradation of heat-treated plasma was studied by the parallel formation of BKE and nonprotein N (NPN) in the digestion mixture. The breakdown of azocasein was estimated as described above.

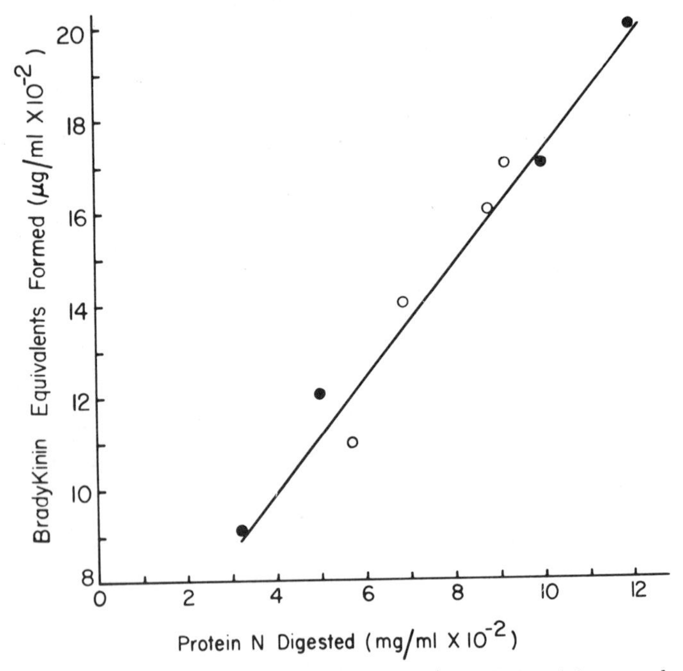

FIG. 16. Correlation between kinin formation and plasma protein digestion by SUT-67 (27°C). ●—●, Enzyme change: data obtained by increasing SUT concentration after 10 min of digestion at 27°C. ○—○, Time change: data obtained by digestion with 0.026 mg N/ml of SUT-67 at variable times.

a. Digestion of plasma proteins. Evidence for the proteolytic activity of SUT-67 was sought by determining whether a relationship existed between the formation of kinin and the digestion of proteins in the reaction mixture. The latter was determined as the increase of NPN estimated on the supernates remaining after precipitation in the presence of 5% trichloracetic acid. The reaction was carried out by treating 4-ml aliquots of HTP substrate containing 10.34 mg protein N/ml with 1-ml portions of SUT-67 so that the final toxin concentrations ranged from 0.007 to 0.1 mg N/ml. The samples were assigned to a variable enzyme or a variable time series at constant enzyme concentration (0.052 mg N/ml). The supernate was tested for NPN

TABLE XVIII

Digestion of 1.3% Azocasein in 0.1 M Borate Buffer (pH 8.4) by Various Sea Urchin Toxin Preparations

Type	mg N/ml [a]	ΔOD_{390}/ hr	Units [b]
Crude toxin			
SUT-67	0.1767	0.316	3.6
Sephadex fractions			
Fr. I	0.0312	0.164	10.51
Fr. II	0.0356	0	-
Fr. III	0.0363	0	-
Ammonium sulfate fractions			
SAS 33%	0.0404	0.158	7.82
SAS 65%	0.0420	0.221	10.52
SAS 100%	0.0670	0.373	11.13
Hydroxylapatite fractions			
HO-A Fr. I	0.0410	0.417	20.31
HO-A Fr. II	0.2192	0.468	4.27
HO-A Fr. III	0.1273	0.013	0.20
Intermediate	0.0173	0.038	4.38
HO-A Fr. IV	0.0277	0.013	0.93
HO-A Fr. V	0.0076	0.015	3.94
HO-A Fr. VI	0.0779	0.021	0.53
60 min blank	0	0.002	-

[a] mg/ml - 0.5 ml quantities used in test.
[b] Unit of potency is expressed as 1 ΔOD_{390}/mg N/hr.

TABLE XIX

Comparison of the Kininolytic and Kininogenic Properties of Sea Urchin Toxin Preparations at 32°C

Toxin preparation		Kininolysis			Kininogenesis	
Type	Concentration (mg N/ml) [a]	Substrate present: bradykinin (μg/ml) (x 10^{-2})	Bradykinin found after digestion (μg/ml) (x 10^{-2})	Bradykinin destroyed (%)	Substrate present: HTP (mg N/ml) [a]	Net BK [b] formed (μg/ml)(x 10^{-2})
SUT-67 (Crude)	0.0530	4	0	100	8.27	3..3
Seph. Fr. I	0.0093	4	0	100	8.27	2.8
Seph. Fr. II	0.0107	4	2.7	32.5	8.27	1.9
Seph. Fr. III	0.0109	4	3.9	–	–	0
0	0	4	3.9	0	8.27	0

[a] Concentration expressed as milligrams of TCA-precipitable N.

[b] Equivalent bradykinin concentration per milliliter as found by bioassay against bradykinin standards corrected for dilution.

concentration by the Nessler-Microkjeldahl method (Feigen et al. 1954). A separate 2.0-ml aliquot of the original digest was dialyzed overnight against an equal volume of 1% NaCl for the estimation of kinin activity on the guinea pig ileum. The results exhibited graphically in Fig. 16 show a close linear correlation between the formation of kinin and NPN.

b. Digestion of azocasein. For practical reasons it was not possible to carry out the entire experiment at a constant concentration of toxin, but since the reactions appeared to be zero order, the enzyme potencies may be compared by correcting the OD_{390} values to 1 mg of precipitable toxin N. These values are entered in the summary in Table XVIII. Of the preparations tested, only two were entirely devoid of protease action. Thus, all the proteolytic activity appeared to be concentrated in Fraction I of the Sephadex series. High potencies were exhibited by the 65 and 100% SAS fractions. In the case of the hydroxylapatite fractions the protease activity was concentrated in Fraction I, and moderate values were found in Fraction II, Fraction ''Intermediate,'' and Fraction V.

2. Attack on Synthetic Substrates

Before dealing with the question of enzyme specificity, it was useful to determine whether the bradykinin-forming activity could be separated from the bradykinin-inactivating property by fractionation.

a. Comparative kininolytic and kininogenic actions. The relationship between the kininogenic and kininolytic properties of the various SUT preparations was studied by incubating either 4×10^{-2} mg/ml of bradykinin or 8.27 mg N/ml of HTP with various Sephadex fractions present in a concentration of 0.01 mg N/ml.

Table XIX shows that the crude toxin and the first Sephadex fraction were the most potent in kininolysis as well as in kininogenesis, and that Fraction I, even though it was present in a concentration only 1/5 as great as that of the parent material, yielded almost as much kinin as did the crude toxin. An inspection of the results obtained with the three fractions further reinforces the association between the net kinin-forming power and the kininolytic action: Fraction II, which destroyed only 32.5% of the BK initially present, formed only 68% as much active material from plasma as did Fraction I, whereas Fraction III, which had no measurable kininolytic effect at the concentration used, produced no active material from the substrate.

b. Inhibition by Trasylol. Trasylol inhibits trypsin and chymotrypsin and has been used to distinguish kinin-liberating bacterial proteases (Nagarse, Clostripaine) from the above since the latter are not affected by this inhibitor. To find out whether the SUT proteins were trypsinlike, their kininolytic activity with respect to bradykinin was studied in the presence and absence of Trasylol. To summarize the results obtained by bioassay, we found that the MED of bradykinin was 4.4×10^{-4} μg/ml. When toxin was present in the absence of inhibitor, the apparent MED (based on the initial concentration of BK in the reaction vessel before digestion) was 5.5×10^{-2} μg/ml, but in the presence of the inhibitor that value dropped to 1.2×10^{-3} μg/ml. Dividing the experimental values by the MED of bradykinin, it is evident that 125 MED's were digested by the noninhibited enzyme, but only 2.73 MED's were degraded in the presence of Trasylol.

c. Proof of chymotrypsin-like specificity. Since it was now clear that the kininolytic action of SUT could be distinguished from the kininogenases and kininases by Trasylol, we next inquired whether the activity of the crude preparation was

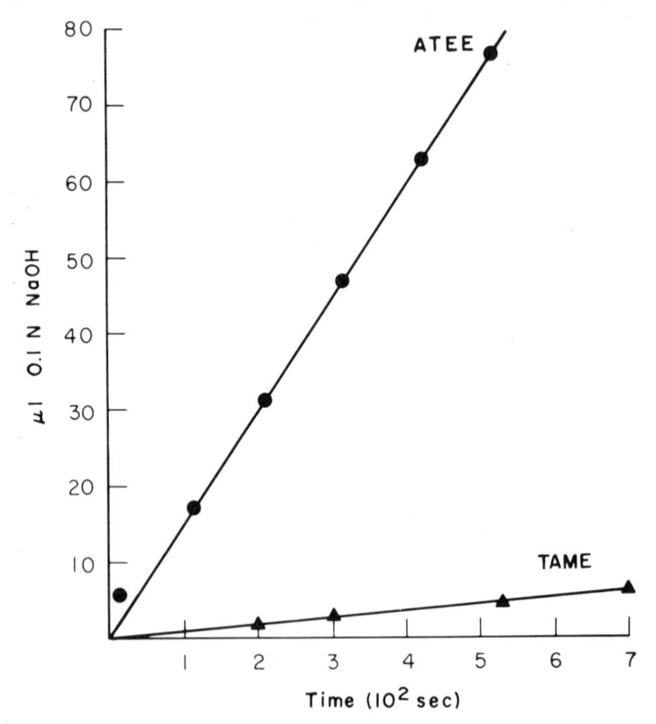

FIG. 17. Time course of hydrolysis of 0.01 M ATEE and 0.01 M TAME by crude SUT-67 (10 μg/ml) at 30°C.

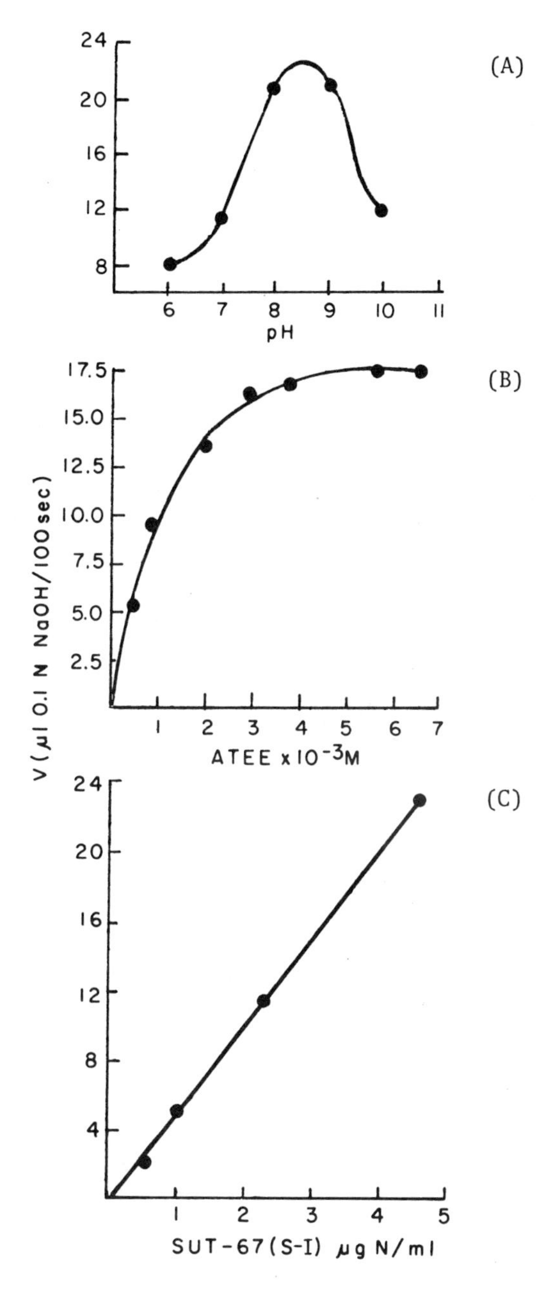

FIG. 18. Effect of pH, substrate concentration, and toxin concentration on the velocity of hydrolysis of ATEE by Sephadex Fraction I (S-I).

predominantly chymotrypsin-like by studying the time course of hydrolysis of 0.01 M TAME and 0.014 M ATEE at 30°C in the presence of 0.0026 mg N/ml of crude toxin. The studies were carried out using the continuous titration described under Methods, page 79. The results displayed in Fig. 17 demonstrate convincingly that the activity of the parent material is chymotrypsin-like.

d. Conformance of SUT preparations to classic enzyme kinetics. A preliminary study of the Sephadex fractions showed that only S-I was able to hydrolyze ATEE, and although the specific hydrolytic activity had been increased over that of the parent material, S-I had only 1/16 of the catalytic power of crystalline alpha chymotrypsin.

Examination of the common functional features of Sephadex Fraction I showed it to behave as a classic enzyme. The function-relating hydrolytic rate to the ambient pH, shown in Fig. 18A, indicates an optimum between pH 8 and pH 9. Figure 18C shows that the dependence of the rate of hydrolysis of 0.014 M ATEE on enzyme concentration is completely linear between 0 and 5 μg N/ml. The dependence of digestion velocity on substrate concentration is exhibited for Fraction I by the hyperbola shown in Fig. 18B for ATEE concentrations ranging from 4.88 x 10^{-4} to 7.43 x 10^{-3} M. The reaction constants evaluated from the usual form of the Lineweaver-Burk plot (Fig.19) gave a value of 26.7 μl of 0.1 N NaOH x 100 sec^{-1} for the maximum velocity and a K_m of 1.9 x 10^{-3} M.

P. Distribution of Chymotrypsin Activity and Its Relationship to Lethal Toxicity and Hemolysis

Table XX shows that only four of the preparations in the entire array had trypsinlike specificity. In the Sephadex system the activity was present only in the first fraction, S-I; but in the hydroxylapatite series it was distributed between HOA I and HOA II. The relative efficiencies of the two of the methods of preparation can be judged from the turnover numbers, i.e., the values entered under V_{max}/mg N. The concentration of activity by Sephadex fractionation was 2.7-fold, whereas the increase in potency obtained by the hydroxylapatite method (HOA II) was 6.7 times greater than that of the crude preparation.

All the preparations were hemolytic with respect to washed rabbit cells. Hydroxylapatite fractionation produced hemolysins of greater potency than the Sephadex separation. Although the

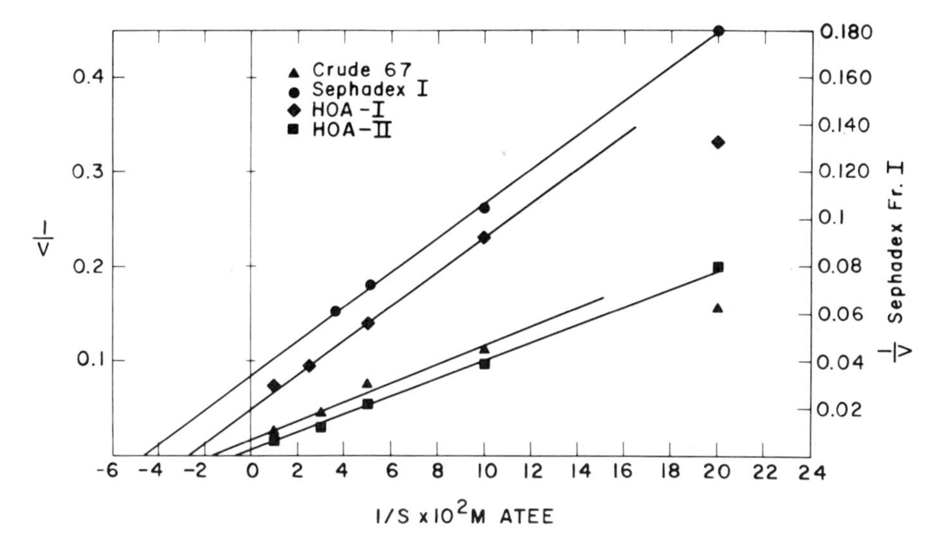

FIG. 19. Lineweaver-Burk plots for the hydrolysis of ATEE by SUT-67 (26 μg/ml), Sephadex Fraction I (4.5 μg/ml), HOA Fraction I (10 μg/ml), and HOA Fraction II (10.9 μg/ml). The scale on the right-hand ordinate refers to the data for Sephadex Fraction I.

most esterolytic preparation, Fraction HOA II, was also the most active hemolysin, this association is not the general case since the most potent esterase of the Sephadex series, Fraction I, was the weakest hemolytically. The association between the caseinolytic and esterolytic power is also equivocal: In the Sephadex series both properties were concentrated to the same degree, but in the HOA series the association was inverted. Since the association between the caseinolytic and hemolytic potencies also varied with the technique of purification, it appears that the three functional properties may be due to separate components.

The next question is that of the relation of catalytic potency and specificity to the lethal toxicity of the preparations studied. The present results show that the lethal toxicity does not depend on the formation of bradykinin from the heated plasma. The most toxic preparation of the Sephadex series (Fraction II) had no measurable esterolytic or proteolytic action, and, additionally, tests in this laboratory have shown that mice can survive intravenous doses of bradykinin at least up to 0.1 mg/20 gm.

TABLE XX

Summary of Functional Activities of Various Sea Urchin Toxin Preparations

Type	Specific toxicity LD_{50} (per mg protein N)	Enzymology: Azocasein (OD_{390}/mg N hr^{-1})	Enzymology: ATEE K_m	Enzymology: ATEE V_{max} [a]	Enzymology: ATEE V_{max}/mg N	Hemolysis H_{50} (mg N) [b] (Rabbit Cells)
Crude Toxin						
SUT-67	7.75×10^3	3.58	5.5×10^{-3}	55.5	213.4	2.0×10^{-5}
Sephadex Fractions						
Fr. I	3.5×10^3	10.51	1.92×10^{-3}	26.7	592.6	3.0×10^{-4}
Fr. II	2.0×10^4	0	no activity	-	no activity	1.4×10^{-5}
Fr. III	1.4×10^e	0	no activity	-	no activity	3.0×10^{-5}
Hydroxylapatite Fractions						
HO-A Fr. I	0	20.31	3.57×10^{-3}	20.0	188.6	2.9×10^{-5}
HO-A Fr. II	1.5×10^4	4.27	1.46×10^{-2}	154.0	1412.8	1.2×10^{-6}
HO-A Fr. III	6.6×10^3	0.20	no activity	-	no activity	9.8×10^{-6}
HO-A Intermediate	2.2×10^3	4.38	no activity	-	no activity	-
HO-A Fr. IV	5.5×10^3	0.93	no activity	-	no activity	1.0×10^{-5}
HO-A Fr. V	0	3.94	no activity	-	no activity	4.6×10^{-5}
HO-A Fr. VI	0	0.53	no activity	-	no activity	7.3×10^{-5}

[a] Microliters 0.1 NaOH/100 sec.
[b] Milligrams protein N to produce 50% lysis.

V. DISCUSSION

The demonstration that certain components of SUT were chymotrypsinlike, whereas none was found to have trypsinlike activity explains the reason for the rapid degradation of the reaction product and the parallel variation of kininogenic and kininolytic activity. Additionally, this finding partially accounts for the variation of the ileum ratios between the SUT and trypsin digests, shown in Table XVI.

Trypsin is particularly active on peptides in which the carboxyl group is given by a lysine or arginine residue, and chymotrypsin is maximally effective when the carboxyl of the peptide linkage is given by tyrosine or phenylalanine. The two enzymes can be functionally distinguished from each other by their selectivity in the hydrolysis of synthetic esters. Thus, trypsin can hydrolyze arginine esters such as TAME but not tyrosine esters like ATEE, whereas chymotrypsin has the opposite specificity. A further distinction, particularly applicable to the present case, is the fact that chymotrypsin is known to be an effective bradykininase whereas trypsin does not destroy bradykinin. The reason for this is that chymotrypsin attacks the bradykinin molecule.

$$\mathrm{H\cdot \overset{1}{Arg}\cdot \overset{2}{Pro}\cdot \overset{3}{Pro}\cdot \overset{4}{Gly}\cdot \overset{5}{Phe}\cdot \overset{6}{Ser}\cdot \overset{7}{Pro}\cdot \overset{8}{Phe}\cdot \overset{9}{Arg}\cdot OH}$$

both at the $\mathrm{\overset{5}{Phe}\cdot\overset{6}{Ser}}$ and the $\mathrm{\overset{8}{Phe}\cdot\overset{9}{Arg}}$ positions, whereas trypsin does not have this specificity. On the other hand, trypsin can degrade the related peptide, kallidin (Met-Lys-bradykinin), by attacking the Lys·Arg bond at the amino terminal end to form bradykinin.

The properties of the active fractions are congruent with these actions of chymotrypsin. We have established that they attack ATEE but not TAME and that their bradykininolytic power, which can be blocked by Trasylol, appears to be correlated with their activity as kininogenases.

VI. RANGE OF ACTION OF SUT COMPONENTS

We have identified and studied in detail three of the principal actions of SUT; histamine liberation, the formation of bradykinin, and hemolysis. Although all these processes are individually toxic - in the usual sense of the term - none of them, singly, can account for the lethal effect of any of the individual fractions. The studies reported in this section were made to assess the range of action of the several fractions on certain specific physiological processes in mammals and in marine invertebrates.

The results of these studies have not yet been published but are available, as government progress reports, from this laboratory. The specific findings summarized in this section concern the effects of these fractions on (a) the process of active transport across the rat stomach, (b) lethal toxicity in crabs, and (c) a survey of their effects on molluscan hearts and the body muscles of marine animals.

A. Active Transport

These studies made in our laboratory by Dr. Vaughan and Dr. Ackerman showed that all fractions except S-II depressed active transport. Addition of S-II to the ''mucosal'' chamber of the Ussing apparatus stimulated the process. This fraction had no esterolytic activity and was not particularly hemolytic. Conversely, the most hemolytically potent fractions had minimal effects on transport.

B. Lethal Toxicity in the Crab

Injection of SUT immediately produced a flaccid paralysis, followed by a loss of the righting reflex, and eventual failure of the grasping reflex.

Lethal toxicities were studied on specimens of the crab *Pachygrapsus crassipes* collected from the rocks in Santa Barbara harbor. The average weight of 107 specimens was 23.6 gm. Test doses of the various preparations were injected directly into the hearts of test animals in volumes not exceeding 0.25 ml and the survival time determined. The mean results of tests on the several hydroxylapatite fractions are presented in Table XXI, which gives the minimal lethal dose per milligram toxin nitrogen.

An inspection of Table XXI shows no correlation to exist between the two assays. In the crab test the activity is concentrated in Fraction $HOAi_2$. The specific lethal toxicity of this fraction, 20,000, is ten times greater than the potency of the neighboring fraction, HOA-IV, as well as that of the crude preparation. Such orders of purification are not seen by the mouse test.

TABLE XXI

Summary of Lethal Toxicities as Determined in Mice and Crabs

Type	Fraction	% Crude toxin	Mouse LD_{50}/ mg N	Crab MLD/ mg N	V_{max} mg N ATEE	S_{20}	Antigenic constitution
Crude SUT-67	–	100.0	7,750	1,923	213.4	20/6/2.1	A, A′, B
Sephadex	I	41.8	3,500	0	592.6	20/6.4	B
	II	27.8	20,000	1,492	0	5.68	A, A′
	III	30.4	1,400	80	0	2.00	Different [a]
Hydroxylapatite [b]	I	49.4	0	10	188.6	2.20	Different [a]
	II	3.3	15,000	128	1,412.6	6.40	Different [a]
	III	1.2	6,600	1,000	0	6.40	–
	i_1	1.5	3,584	862	0	–	B
	VIII(i)	26.2	2,200	833	0	6.50	–
	i_2	5.8	8,333	20,000	0	–	B
	IV	2.2	5,500	2,083	0	–	–
	V	10.2	0	66	0	1.80	–
	VI	1.0	0	0	0	–	–

[a] Unidentified components (pending resolution with specific anti-HOA sera).
[b] In order of elution.

C. Action on Clam Heart

Positive inotropic reactions were elicited in the isolated heart of Tivela stultorum to serotonin and noradrenaline. Acetylcholine depressed the beat and histamine was without effect. SUT in effective doses increased the tone of failing hearts and tended to arrest them in systole. Fraction HOA-VI, given to a failing heart on the second day of maintenance in the muscle chamber, reestablished a condition of pulsus alternans into a regular rhythm of beat at the initial control rate (Fig. 20).

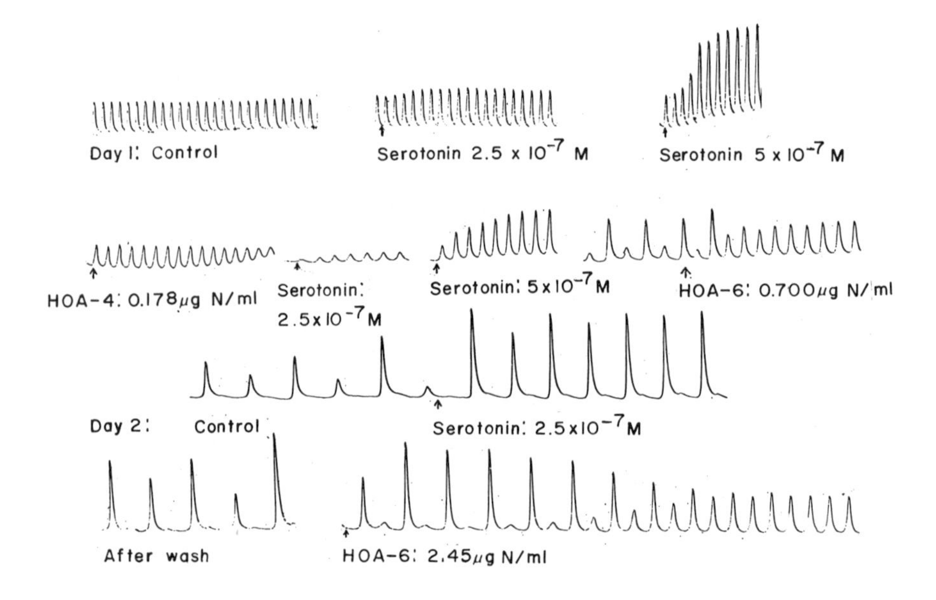

FIG. 20. Chronotropic effects of SUT fractions HOA-IV and HOA-VI tested on clam heart over a 2-day period. The heart was maintained in seawater at 18°C during testing, but was stored overnight in 4°C.

VII. IDENTIFICATION OF SUT COMPONENTS BY IMMUNOELECTROPHORESIS

From the plethora of actions of sea urchin toxin described in this study we have shown that at least three of them (histamine release, bradykinin formation, and lethal toxicity) can be ascribed to the specific actions of different protein molecules.

To determine whether this is indeed the case, and particularly to assess whether the enzymatic activity could be separated from either the mouse or crab lethality, we studied the pattern of cross reactions shown by the several fractions to antisera raised in rabbits against the formalinized toxoids.

Accordingly, anti-SUT-67, anti-S-I, anti-S-II, and anti-S-III were prepared for the immunoelectrophoretic work. This was carried out in 1% barbital buffer, pH 8.2 at an ionic strength of 0.5 μ.

A. Characterization of SUT-67 with Homologous Antisera

The parent toxin contains three antigens: A, A′ and B. A and A′ are similar in charge and molecular size, whereas B moves more slowly and is identifiable only with "later" antiserum.

1. Toxic Component in Mice

(a) Fraction S-I contains only antigen B which can be resolved into two components of the same charge but of differing molecular weight by specific antiserum. (b) S-II consists of A

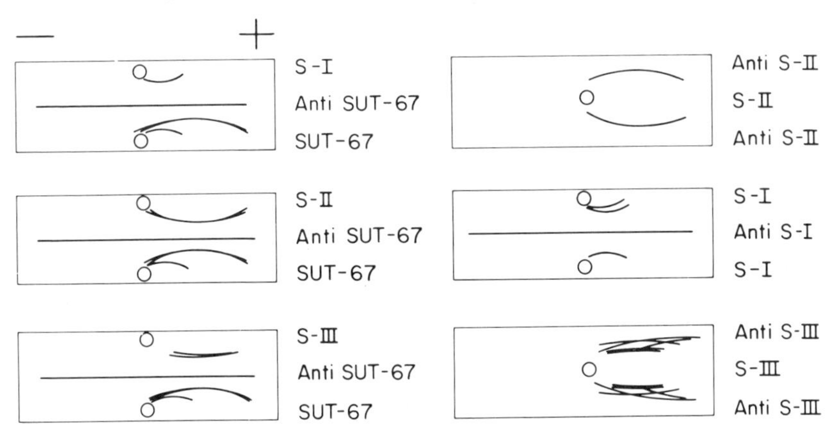

FIG. 21. Antigenic constitution of protein components in pedicellarial toxin of sea urchin (SUT-67). Electropherograms showing relationships between the various Sephadex fractions and the crude toxin are given in the left-hand column. The antiserum used was made against the parent preparation, SUT-67. The right-hand column shows antigenic components present in the individual fractions, as revealed by specific antisera.

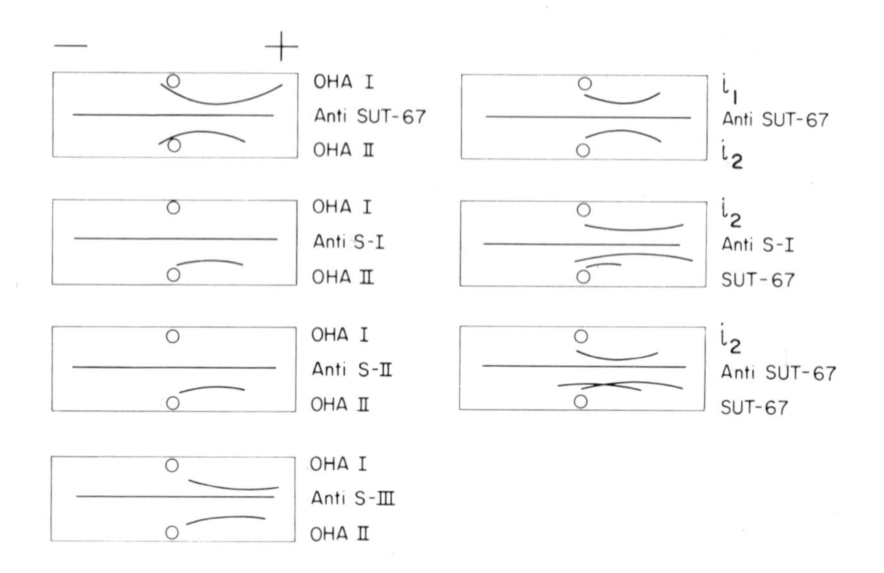

FIG. 22. Immunoelectrophoretic identification of proteolytic, mouse-toxic, and crab-toxic components in hydroxylapatite fractions of sea urchin toxin. After electrophoresis these fractions were tested with various antisera raised against either the crude material or its Sephadex fractions.

and A´, but primarily of A as revealed by the specific antiserum. (c) S–III is a separate set of antigens and consists of multiple components (Fig.21).

Since S–II is devoid of enzymatic activity and contains only antigen A, it is probable that this antigen is the toxic moiety. This antigen is predominant in the parent preparation (SUT–67) and is present above in the S–II and anti-S–II system.

2. Chymotrypsin Component

Since none of the Sephadex fractions showed toxicity in exclusion of other actions, we searched for the enzymatic components in the HOA preparations.

HOA–I is the only member of that series having chymotrypsin activity and virtually no lethality for crabs. Tests against anti-SUT–67 and anti-S–III show that it is a single antigenic determinant (Fig.22). It is highly specific and does not cross react with either anti-S–I or S–II. This antigen is also identifiable in the parent toxin and in its Sephadex fractions.

HOA-II is also a single antigen and reacts with all four antisera. HOA-II contains both functional activities and, since it is a single antigen, it may be bifunctional.

3. Toxic Component Active on Crabs

Fraction $HOAi_2$ has the highest specific lethal activity in crabs: 2×10^4 MLD/mg N. Studies with the A antisera reveal that only a single antigenic component for this preparation exists and that it is comparable to antigen B of the Sephadex series.

VIII. CONCLUSIONS

Hydroxylapatite fractionation produced antigenically pure components. By means of this technique we were able to show the following.

1. The chymotrypsin activity is represented by a single antigen in HOA-I. This fraction is atoxic, it has moderate esterolytic power but is highly potent against azocasein, and it is one of the smaller molecules in this distribution, having a sedimentation constant $S_{20} = 2.2$.

2. The purified fractions having the greatest esterolytic potencies in their respective systems, S-I and HOA-II, also have high mouse lethal potency. These preparations are electrophoretically single antigens but their common sedimentation constant is 6.4. (Thus the smaller enzymatic component may be a subunit.)

3. The toxic activity also appears as a pure component in other toxic fractions: S-II ($S_{20} = 5.68$), HOAi, and $HOAi_2$.

The fact that substances that are ''toxic'' to one species of animal may be found to be atoxic in another is not unusual. When these toxins are ensembles of enzymes, as in the case of sea urchin toxin, this phenomenon may have various explanations. For example, the recipient may lack the proper substrate (or suitable activator); there may be no suitable target cells for the products of the reaction; and the enzyme may be exquisitely temperature dependent so that the administration of a substance to a poikilotherm may be ineffective because the system at interest lies outside of the effective temperature range of the reaction. There are many other probable explanations such as the presence of natural inhibitors in the tissues of the recipient. Differences in complexity of the two species of animal are obviously a factor of paramount importance. The greater the number of control and regulatory sites in one respect may make a mammal more susceptible to the effects of a given preparation containing a variety of components, thus leading to an apparently broader distribution of toxicity among the fractions and to an apparently lower degree of purification as measured in a bioassay.

The studies summarized in this report are based in part on work performed in this laboratory. For further information see Alender et al. (1965) and Feigen et al. (1966, 1968, 1970a,b).

ACKNOWLEDGMENTS

The past support of the National Science Foundation (GB-16364), the Office of Naval Research (Biochemistry Branch) (N0014-67-A-0112-0040), the National Institutes of Health (GMAI-13942) and U.S.P.H. (Allergy & Immunology Branch) (Grant No. USPH 5-T01-AI00327-03) is gratefully acknowledged. In addition, the authors wish to thank the following colleagues for their contributions made during the past 6 years to this effort: Professor Charles Alender, Dr. Joseph T. Tomita, Father E. Sanz, O.P., Dr. Roger A. Pfeffer, and Dr. Gabor Markus. Particular thanks are due to Mr. Norman Lammers of the Santa Barbara Marine Station for his hospitality and seamanship. To Dean Terrence A. Rogers, University of Hawaii Medical School, go special thanks for making arrangements and providing the necessary facilities at the Pacific Biomedical Research Center, University of Hawaii, which immeasurably facilitated the collection and processing of the material.

REFERENCES

Alender, C.B., Feigen, G.A., Tomita, J.T., Toxicon 3, 9 (1965).
Brody, S., "Bioenergetics and Growth." Reinhold, New York, 1945.
Charney, J., and Tomarelli, R.M., J. Biol. Chem. 171, 501 (1947).
Feigen, G.A., Campbell, J.M., Sutherland, G.B., and Markus, G. J. Appl. Physiol. 7, 154 (1954).
Feigen, G.A., and Nielsen, C., Science 154, 676 (1966).
Feigen, G.A., Hadji, L., Pfeffer, R.A., and Markus, G., Physiol. Chem. Phys. 2, 309 (1970a).
Feigen, G.A., Hadji, L., Pfeffer, R.A., and Markus, G., Physiol. Chem. Phys. 2, 427 (1970b).
Feigen, G.A., Sanz, E., and Alender, C.B., Toxicon 4, 161 (1966).

Feigen, G.A., Sanz, E., Tomita, J.T., and Alender, C.B., Toxicon 6, 17 (1968).
Frieden, C., J. Biol. Chem. 239, 3522 (1964).
Jacobsen, S., Nature (London) 210, 98 (1966).
Kellaway, C.H., and Trethewie, E.R., Quart. J. Exptl. Physiol. 30, 121 (1940).
Kendall, F.E., J. Clin. Invest. 16, 921 (1937).
Lotka, A.J., "Elements of Mathematical Biology." Dover, New York, 1956.
Markus, G., McClintock, D.K., and Castellani, B. A., J. Biol. Chem. 242, 4395 (1967).
Markus, G., and Werkheiser, W.C., J. Biol. Chem. 239, 2637 (1964).
Pennell, R.B., in "The Plasma Proteins" (F.W. Putnam, ed.) Academic Press, New York, 1960.
Prado, J.L., Acta Physiologica Latino Americana 14, 215 (1964).
Prado, J.L., Mendes, J., and Rosa, R.C., Anais da Acad. Brasileira de Ciéncias 37, 295 (1965).
Rocha e Silva, M., Beraldo, W.T., and Rosenfeld, G., Amer. J. Physiol. 156, 261 (1949).
Shore, P.A., Burkhalter, A., and Cohn, Jr., G.H., J. Pharmacol. Exptl. Therap. 127, 182 (1959).
Trapani, I.L., and Feigen, G.A., J. Infect. Dis. 104, 261 (1959).
Tu, A.T., in "Neuropoisons: Poisons of Animal Origin" (L.L. Simpson, Ed.), Vol. 1. Plenum Press, New York, 1971.
Werle, E., and Berek, U., Biochem. Z. 320, 136 (1950).

Chapter 4

ULTRASTRUCTURE OF PHYSALIA NEMATOCYSTS

William H. Hulet*
Division of Undersea Medicine
School of Medicine, University of Miami
Miami, Florida

J. L. Belleme and G. Musil
Electron Microscopy Laboratory
Veterans Administration Hospital, Miami, Florida

Charles E. Lane
Rosenstiel School of Marine and Atmospheric Science
University of Miami, Miami, Florida

The siphonophore *Physalia physalis* or Portuguese-man-of war is a floating colonial hydrozoan that is particularly well known to bathers along the western Atlantic coast. Inadvertent contact with the venomous fishing tentacles produces a severe and painful sting that may last for several days. For many organisms, blundering into the tentacles of *Physalia* is a fatal mistake since the trailing retractable tentacles serve as an effective means of food capture. The length of each fishing tentacle is studded with "batteries" of nematocyst-containing cells. These cells, often referred to as cnidoblasts, are located in the epidermis (Lane and Dodge 1958). It is the nematocyst, a cell organelle, that contains the toxin.

Nematocysts are characteristic of all coelenterates, and Weill (1934) has described more than a dozen taxonomically useful categories. *Physalia physalis* has only one type of nematocyst. Basically it consists of a barbed thread coiled within a spherical capsule. Hyman (1940) states that all nematocysts discharge by eversion of the thread, but as recently as 1961, Chapman and other investigators (1961) could not accept the idea that the heavily armed thread in the stenotelic nematocyst of *Hydra* leaves the capsule by turning inside out. In a review

*Present address: Marine Medicine Division, Marine Biomedical Institute, University of Texas, Galveston, Texas

of research on nematocysts, Picken and Skaer (1966) present data on the geometry of discharge in nematocysts of the sea anemone Corynactis viridis. Their electron micrographs clearly show that the thread everts on discharge.

In continuing our work on Physalia and its toxin, we planned to examine the nematocyst with the electron microscope and, as a first step, to clarify the mechanics of discharge. To our great dismay, countless attempts to section a nematocyst failed. The contents of the capsule shattered and were lost from the section. We had no difficulty in preparing sea anemones and corals for electron microscopy, but Physalia eluded our efforts for many months. In the remainder of this chapter we will show how we solved the problem of tissue preparation and demonstrate the pertinent features of nematocyst discharge in Physalia.

I. MATERIAL

For preparation of undamaged tentacles, we collected live Physalia from the Straits of Florida while on board the R/V Gerda. Free undischarged nematocysts were obtained from fishing tentacles allowed to autolyze in the cold for 24 to 48 hr. Lane and Dodge (1958) previously described the details of this procedure.

II. METHODS

A. Light Microscopy

The addition of a few drops of Chlorox to a suspension of nematocysts is a simple method to induce discharge. For preparation of stained permanent mounts we used two procedures. In the first, intact and discharged nematocysts were fixed in 4% formaldehyde for 1 hr and then stained with 1% aqueous eosin solution for 30 min at 100°C. After cooling, the nematocysts were washed in 95% ethanol for 5 min and resuspended in absolute ethanol. The alcohol was replaced with xylene and the preparation mounted on a glass slide. In the second method, we fixed the nematocysts in 2.5% phosphate-buffered glutaraldehyde for 1 hr. After washing with two changes of 0.1 M neutral phosphate buffer, we "stained" with 1.33% buffered osmium tetroxide solution. We dehydrated with ethanol and suspended the nematocysts in xylene prior to mounting.

B. Electron Microscopy

Immediately after collection, we immersed small pieces of Physalia tentacles in a fixing solution composed of glutaraldehyde, acrolein, paraformaldehyde, and s-collidine buffer. According to the directions given by Hayat (1970), the tissue was postfixed with 2% osmium tetroxide. We used acetone for dehydration prior to embedding the tissue in plastic. Isolated nematocysts were fixed in 2.5% phosphate-buffered glutaraldehyde for 8 hr. We maintained the osmolality of all fixatives and phosphate buffer washes at 2000 milliosmols per liter by the addition of sodium chloride. All aqueous solutions contained 0.1 M phosphate buffer at pH 7.2. The entire procedure was carried out in the cold (4°C). Osmium tetroxide, 1.33%, was used for postfixation (1 hr). We dehydrated the fixed nematocysts with ethanol and gradually substituted propylene oxide for optimum miscibility with the embedding plastic. Osmium penetrated the capsule of undischarged nematocysts only when the sample was pretreated with an aldehyde fixative, either glutaraldehyde or formaldehyde.

We believe the crucial factor that enabled us to cut thin sections of Physalia nematocysts was the use of a low-viscosity embedding medium. The Spurr epoxy embedding method is based on vinyl cyclohexene dioxide, which readily penetrated hard tissues and even mineral rocks (Spurr, 1969). Thin sections were cut with a diamond knife on a MT-2 ultramicrotome, stained with uranyl acetate and lead citrate, and examined in a Philips EM 300 transmission electron microscope.

C. Scanning Microscopy

For scanning electron microscopy (Cambridge SEM 4), osmium-fixed nematocysts were washed with distilled water, immersed in liquid nitrogen, freeze dried for 12 hr at -60°C, and gold coated in a vacuum evaporator.

We recently had the opportunity to do a sample run on the Quantimet 720 Image Analysing Computer. In this experiment we measured the diameter of several thousand free nematocysts of P. physalis.

III. RESULTS AND DISCUSSION

The Physalia nematocyst is a barbed thread coiled within a thick spherical capsule. Isolated nematocysts as well as those observed in sections of fishing tentacles are of one structural type and two sizes (Fig.1). From a size analysis with the Image Analysing Computer we obtained a bimodal distribution. With a sample size of 4670 nematocysts, two groups

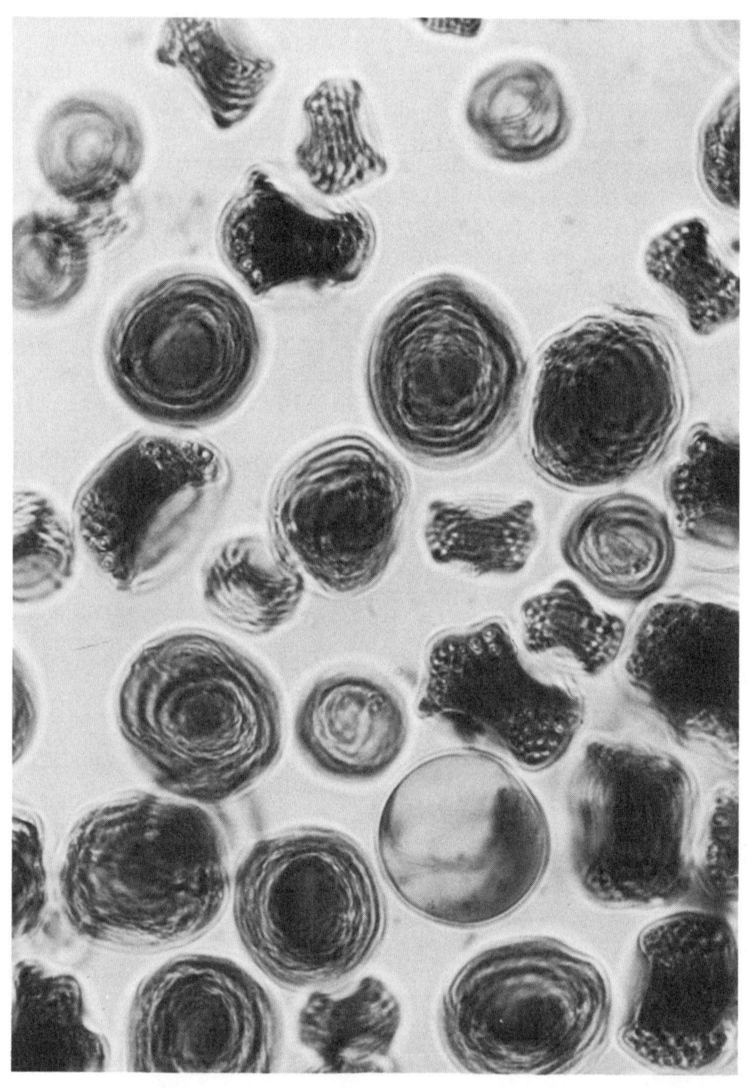

Fig.1. Light micrograph of isolated nematocysts of Physalia physalis; fixation with glutaraldehyde and osmium tetroxide. (X865.)

accounted for 78% of the measurements. Mean diameters were 11.3 μm for the smaller nematocysts (19%) and 25.3 μm for the larger (59%). These data agree closely with the earlier report of Lane and Dodge (1958). Six percent of the nematocysts in our measurements were over 40 μm in diameter. This group probably represents discharged nematocysts in the sample at the time of scanning measurements. Although coelenterate nematocysts are cellular organelles that develop in the cytoplasm, we did not find early stages or what could be called immature nematocysts in histological sections of *Physalia* tentacles. Slautterback (1961) described the differentiation of nematocysts in *Hydra*, and we have seen different stages of development in tentacles of the sea anemone *Condylactis*. We did find in our image analysis study that 435 out of 4670 measurements gave a diameter less than 6.2 μm. This size group may have been immature nematocysts that we did not detect microscopically.

A simple but conceptually valid way to visualize the mechanics of a discharging nematocyst is the process of turning a rubber surgical glove inside out. After stuffing the parts covering the palm and fingers through the opening at the wrist end, air can be trapped inside and the glove held closed. As the partially inflated glove is gently compressed, the fingers begin to evert. It is important to note that eversion begins at the base of the inverted fingers. With additional pressure the everted-inverted junction travels distally until the entire finger is everted. Except for some complexities of rotational symmetry, which we will discuss later, this is the way the nematocyst thread gets out of the capsule.

In our discussion we use the term "mechanics of nematocyst discharge" to describe a sequence of anatomical events. The precise mechanism that incites a living *Physalia* to discharge its nematocysts remains unknown. Sensory hairs (cnidocils) and a nerve network have been described for the cnidoblasts of other coelenterates (Westfall, 1970). If cnidocils are present in *Physalia*, they are not a prominent feature since we have yet to find one directly associated with a cnidoblast. In isolated nematocysts of *Corynactis*, Picken and Skaer (1966) found the intracapsular fluid to have a high osmolality, about 3000 mOsm/liter. They suggest that some mechanism increases the permeability of the capsular wall to water and the rapid expansion of intracapsular fluid exerts a strong driving force for expulsion of the thread.

Returning to *Physalia*, the photomicrographs of Fig. 2 demonstrate several pertinent structural details. As it so often happens, after we reviewed a number of electron micrographs we were able to recognize anatomical details by light microscopy, which previously went unnoticed. Beneath the operculum (Fig. 2a) the coiled thread is anchored to the internal wall of the capsule. An end-on view of several coils is shown in Fig. 2b.

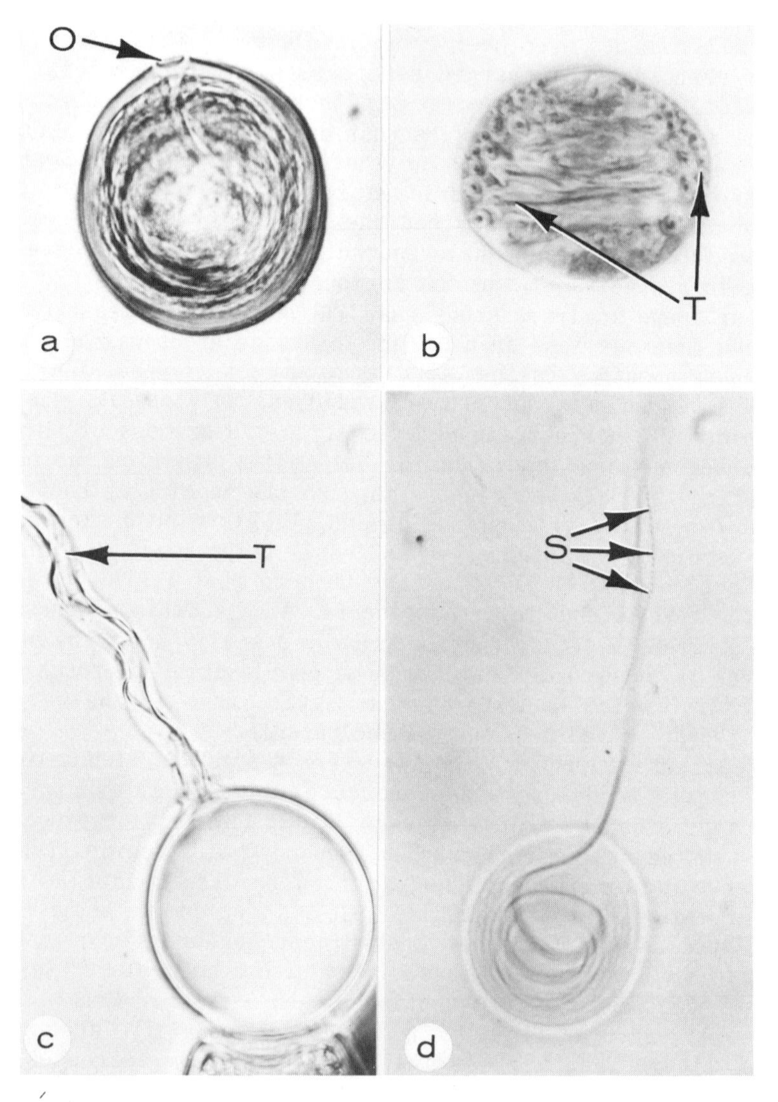

FIG. 2. (a) Undischarged nematocyst showing opercular opening (O) and coiling of the thread; (b) dark area indicates spines inside the thread (T); (c) a discharged nematocyst; (d) partially discharged nematocyst showing helical row of spines (S); eosin stained. (X1200.)

Closely packed spines appear to occlude the lumen of the thread. After ejection of the operculum, the thread begins the eversion process through the opening in the capsule. The thread is attached all around the rim of the opercular opening. As the thread turns inside out, the spine-bearing luminal surface on the inverted thread now becomes the external surface and the spines project to the outside. Figure 2d shows a partially discharged nematocyst. That part of the thread already everted bears rows of spines and serves as a sheath for the advancing remainder of the thread, which has yet to evert. The thread of a completely discharged nematocyst (Fig.2c) spirals away from the capsule in a clockwise rotation. Spines run the length of the thread and since the latter is of constant diameter the *Physalia* nematocyst can be classified as a holotrichous isorhiza (Hand, 1961).

The discharged thread of *Physalia* has three longitudinal pleats and three rows of spines that spiral to the right going away from the capsule (Figs. 3 and 4). The scanning electron micrograph in Fig.4 shows how the spatulate spines overlap and point backward toward the proximal end of the thread. The nematocyst in this picture was arrested in discharge, and in the upper portion one can see a spine coming over the ridge that is the moving junction between the everted and inverted segments of the thread. The structure of the thread is remarkably suited for rapid eversion with minimal frictional resistance. Since the everted thread has right-handed helical pleats, the uneverted thread, as it lies in the capsule or is sliding outward through the sheath formed by the everted portion, must have the shape of a left-handed screw. This engineering feat of *Physalia* and other coelenterates is demonstrated in the transmission electron micrographs of Figs.5 and 6. The unfired thread has the shape of the triquetrum or triskelion, a three-legged ancient magic symbol that runs counterclockwise. This symbol represents one of the simplest forms of rotational symmetry (Weyl, 1952). It should be noted that in sections cut through an intact nematocyst half the coils will rotate to the left and the other half to the right. The spines of the three helical rows form the center of the undischarged thread (Fig.7). They are triangular in shape and of equal size. The reason for the apparent size difference in Fig.7 is because the single plane of the section cuts through different relative levels of each helix of spines. The relative position of the spines can be better understood from a longitudinal section of the thread (Fig.8). Figure 9 shows the cross-sectional appearance of two partly discharged threads. The internal thread is propelled

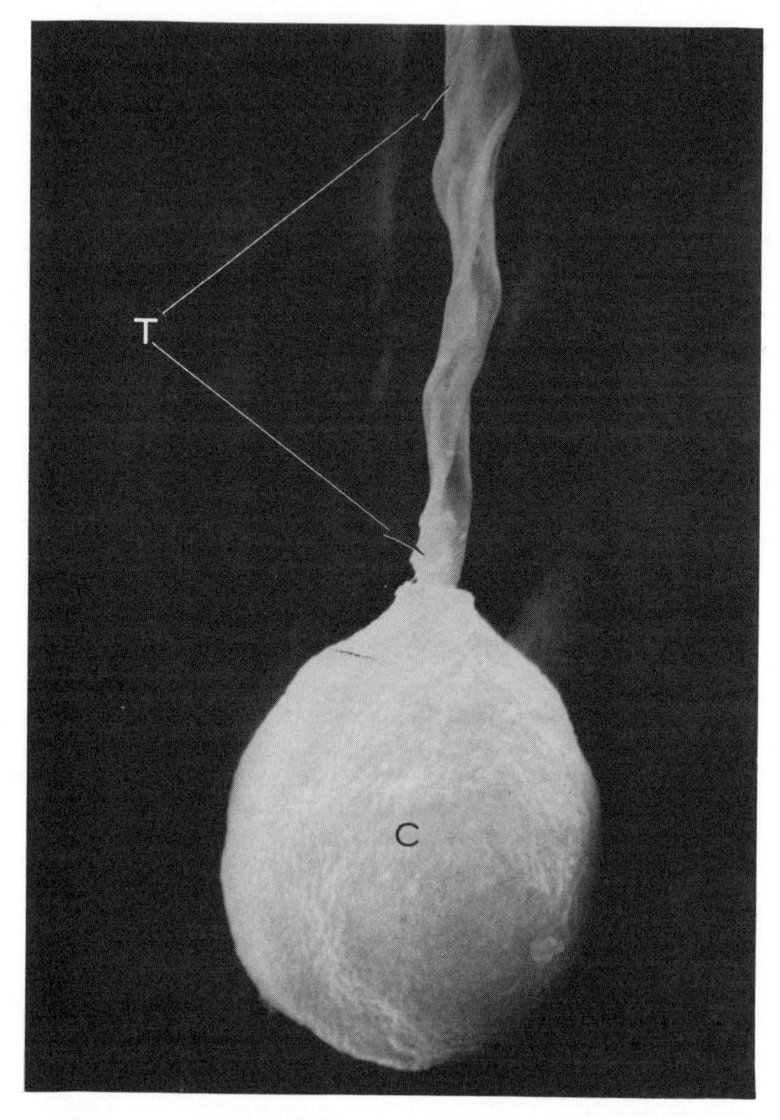

FIG. 3. Scanning electron micrograph showing capsule (C) and thread (T) of a discharged nematocyst. (X3800.)

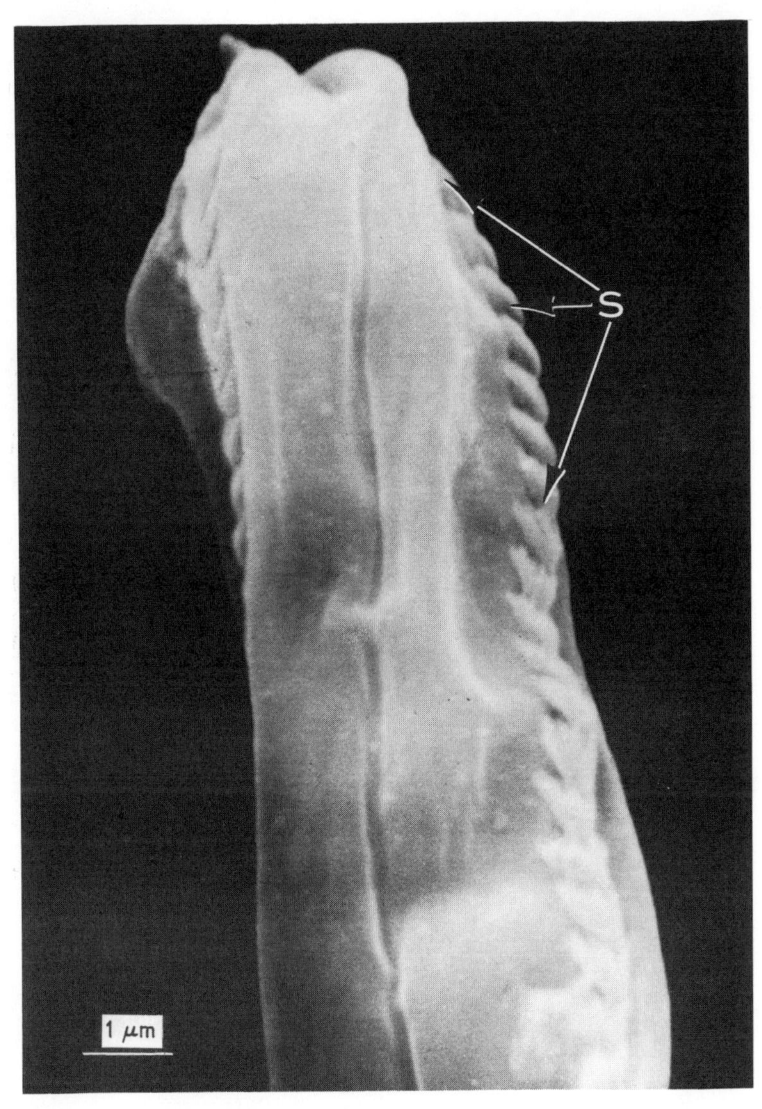

FIG. 4. Scanning electron micrograph demonstrating the spatulate spines (S) of a nematocyst arrested in discharge. (X14,400.)

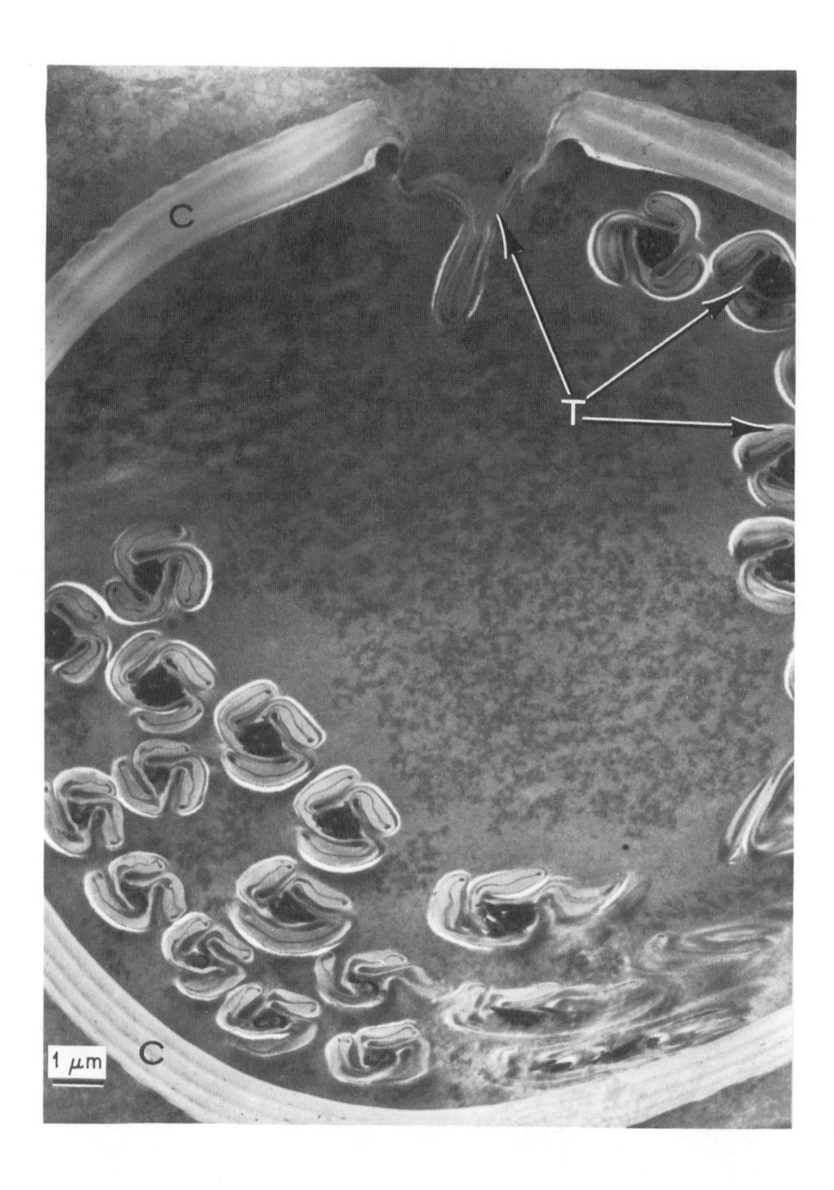

FIG.5. Transmission electron micrograph of a nematocyst *in situ* in a fishing tentacle. Note the layering of the capsule (C) and the triquetrum-shaped thread (T). (X9300.)

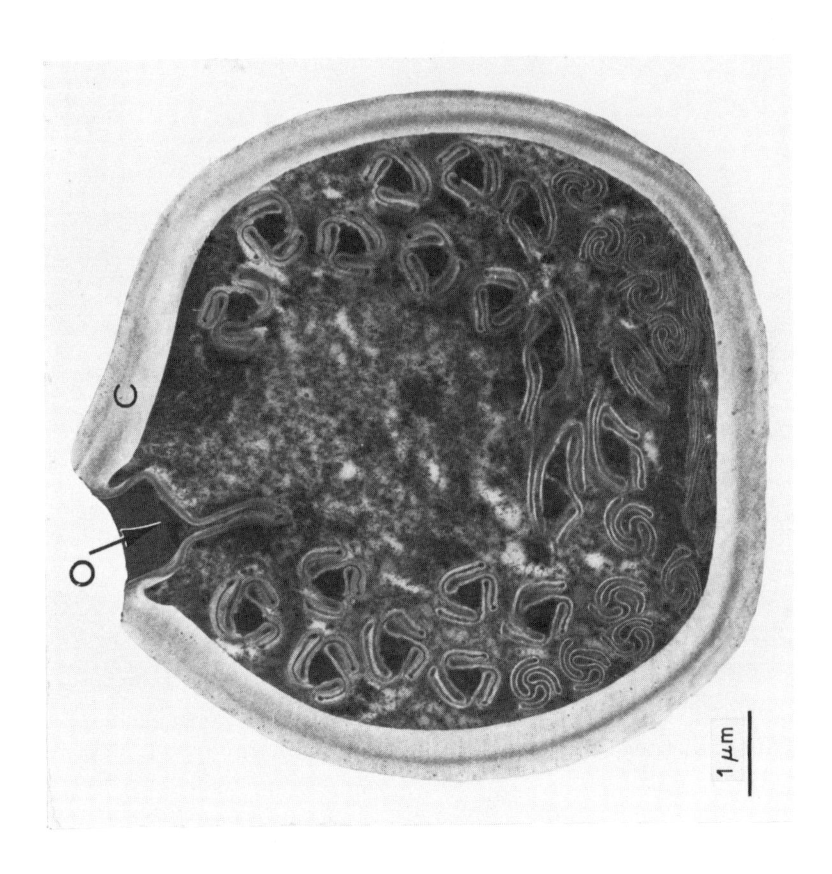

FIG.6. The first part of the thread to evert is located beneath the operculum (O). (X14,400.)

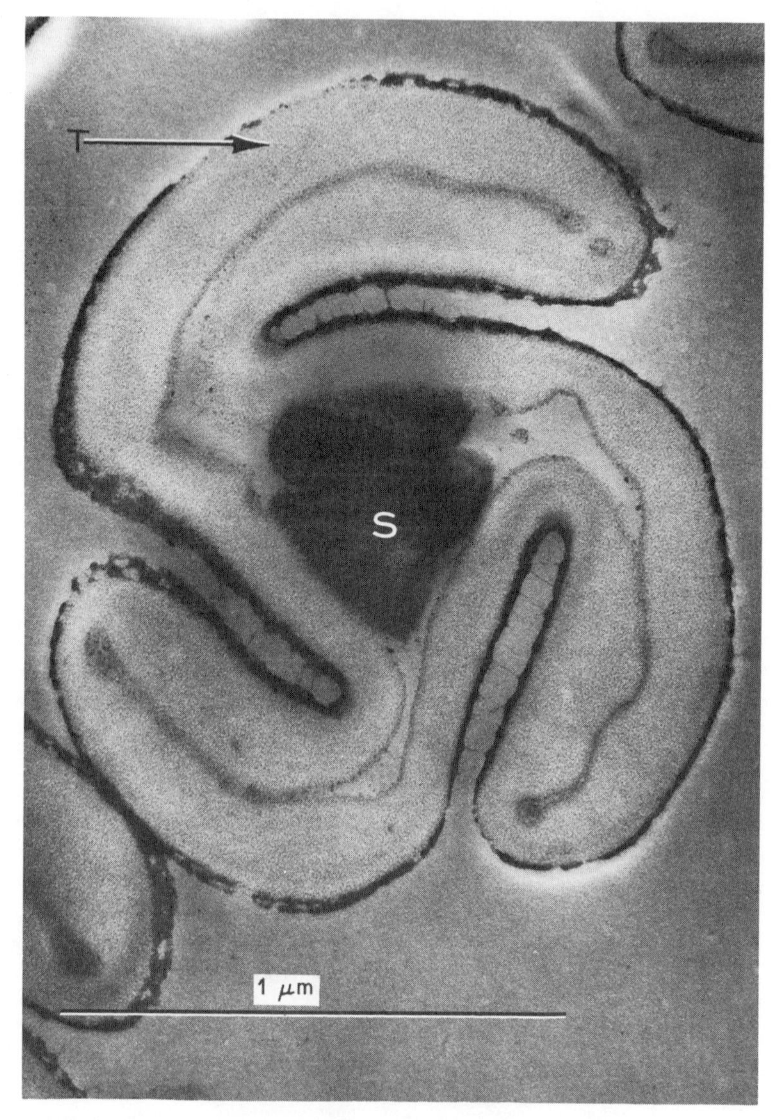

FIG. 7. Cross section of an undischarged thread. Portions of three spines (S) are present on the internal surface. (X84,300.)

FIG. 8. Tangential section of two thread coils. When the thread is discharged the spines (S) will form three external helical rows. (X41,500.)

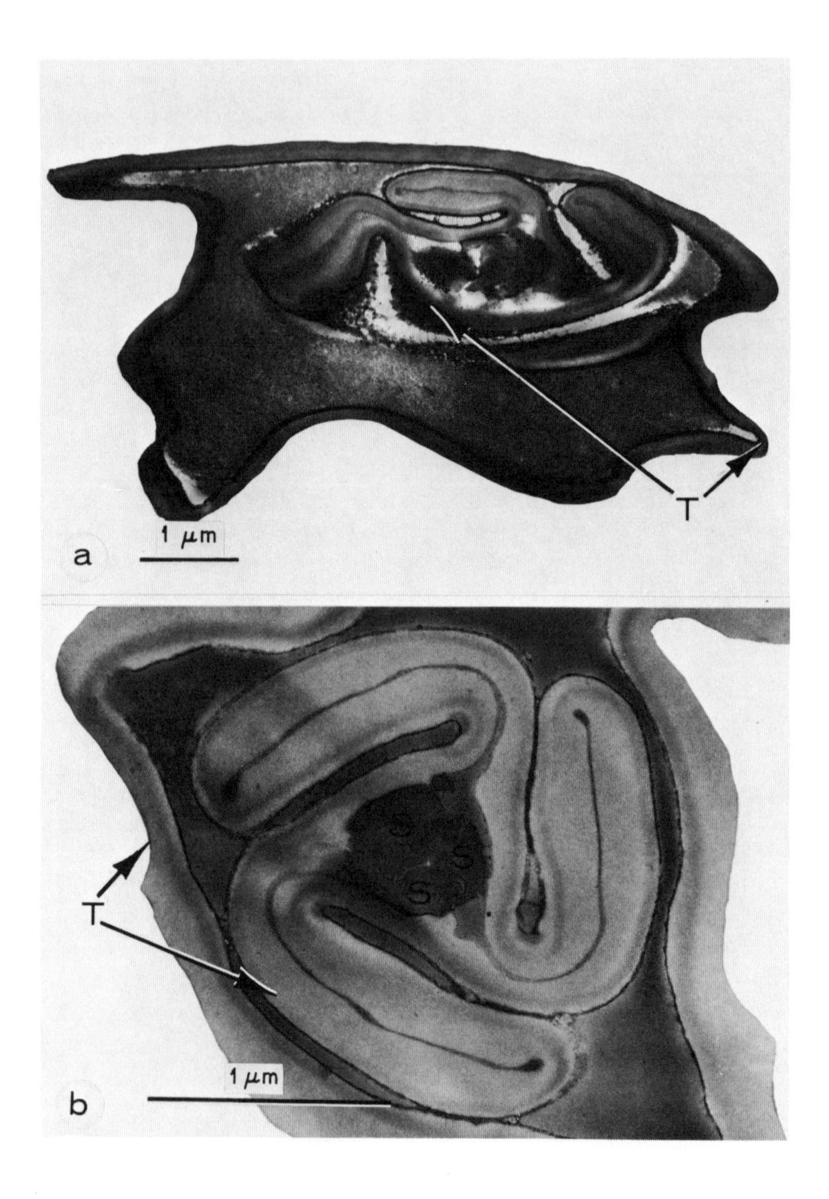

FIG. 9. Section outside the capsule through two partially discharged threads. The uneverted segment is sheathed by the external everted portion of the thread. [(a) X17,600; (b) X41,000.]

through the external everted portion until eversion is complete and all the spines are outside.

We have made one rather obvious conclusion from this study on mechanics of discharge. The barbed thread is in a geometrical position to penetrate and affix itself to the victim's tissue immediately after eversion begins. Total discharge of the nematocyst is not necessary, at least not necessary to produce mechanical injury. The location of the toxin is another question of some importance. We already know that ground up *Physalia* nematocysts yield a highly poisonous fluid, which is not a constituent of the capsule wall. Some of the toxic protein complex may be deposited on both sides of the thread membrane, and, if so, contact with any portion of an ejected thread would be poisonous.

ACKNOWLEDGMENTS

This work was supported in part by Designated Part I Funds, Veterans Administration Central Office, Washington, D.C. We are grateful to Dr. Eugene Small, Department of Zoology, University of Illinois, for the use of the Cambridge Scanning Electron Microscope and to John Jeffery of Image Analysing Computers Ltd. for the analysis of nematocyst size.

REFERENCES

Chapman, G.B., "The Biology of Hydra," p. 131. University of Miami Press, Coral Gables, Florida, 1961.

Hand, C., "The Biology of Hydra," p. 187. University of Miami Press, Coral Gables, Florida, 1961.

Hayat, M.A., "Principles and Techniques of Electron Microscopy," p. 339, Van Nostrand Reinhold, New York, 1970.

Hyman, L.H., "The Invertebrates: Protozoa through Ctenophora," p. 382. McGraw-Hill, New York, 1940.

Lane, C.E., and Dodge, E., Biol. Bull., Woods Hole 115, 219 (1958).

Picken, L.E.R., and Skaer, R.J., "The Cnidaria and Their Evolution," p. 19, Zool. Soc. of London Symposium No. 16, Academic Press, London, 1966.

Slautterback, D.B., "The Biology of Hydra," p. 77. University of Miami Press, Coral Gables, Florida, 1961.

Spurr, A.R., J. Ultrastruct. Res. 26, 31 (1969).
Weill, R., Trav. Stat. Zool. d. Wimereux. Tome 10, 11 (1934).
Westfall, J.A., Z. Zellforsch. 110, 457 (1970).
Weyl, H., "Symmetry," Princeton University Press, Princeton, New Jersey, 1952.

Chapter 5

DIFFERENCES BETWEEN THE ACTIONS OF TETRODOTOXIN AND SAXITOXIN

C. Y. Kao

Department of Pharmacology
State University of New York
Downstate Medical Center
Brooklyn, New York

I. INTRODUCTION

When Dr. Lane first wrote me about this symposium he assigned me the task of discussing the actions of tetrodotoxin, ''differentiating it insofar as possible from saxitoxin.'' To discuss differences between the actions of saxitoxin and tetrodotoxin properly, it is important first to emphasize how similar their actions are. Since the toxins are chemically different and since their fundamental actions on the gating of sodium channels in many excitable membranes are unique, the few and seemingly minor differences in their actions actually may be clues to important membrane groups with which each of these two toxins reacts differently. Because of their similar actions, I propose to discuss these toxins in an intermingled fashion and to bring forth the differences wherever they may arise. Both the cellular and systemic actions are covered as a fairly rational understanding of the diverse and complex systemic effects can be derived from a knowledge of the actions at the cellular levels. Relevant details and references not specifically mentioned here can be found in some review articles (Kao, 1966, 1972; Moore and Narahashi, 1967).

Tetrodotoxin and saxitoxin are among the deadliest poisons known to man. The minimum lethal dose, as determined in the mouse, is 8 μg/kg, which might be contrasted with a dose of 10 mg/kg for the much feared sodium cyanide (Mosher et al.,

1964). Tetrodotoxin is present in the largest quantities in the gonads of some Tetraodontidae fish, some species of which are responsible for clinical cases of tetrodotoxin poisoning. It is also found in some newts of the family Salamandridae, in which several species of the genus *Taricha*, native to California, are the best studied (see Mosher et al., 1964). A toxic material from *Taricha torosa*, at first named tarichatoxin, has now been identified with tetrodotoxin. Saxitoxin is the first of several toxic materials extracted and purified from various types of shellfish, which are responsible for paralytic shellfish poisoning. It is so named because it is extracted from the Alaska butterclam, *Saxidomus giganteus*. All these substances are now known to be chemically identical and have been thought for some years to be derived from some species of a dinoflagellate alga, *Gonyaulax* (Schantz, 1960). Recently, a gonyaulax toxin was purified from an axenic culture of the alga, *Gonyaulax catanella*, and this toxin has been shown to be chemically identical with saxitoxin (Schantz et al., 1966).

Chemically, tetrodotoxin and saxitoxin are different substances. Tetrodotoxin is an aminoperhydroquinazoline compound with a known structure. It has a guanidinium moiety which forms a zwitter ion with the -OH on C-10, a process which is associated with a pK_a of 8.2 for the molecule in solution. At physiological pH, tetrodotoxin is mostly in the cationic form. It also has a hemilactal link between C-5 and C-10, which is not known to exist in any other substance. The structure of saxitoxin has now been revealed (Wong et al., 1971). The elemental formula is $C_{10}H_{15}N_7O_3$, which differs by 1 mole of water of solvation from some previous versions, $C_{10}H_{17}N_7O_4$ (see Kao, 1966, p.1036). The toxin has two guanidinium residues and a carbamate group that links C-6 and C-7 in the form of a cyclol. The -OH group of the carbamate accounts for one of the 2 pK_a's of the compound at 8.24. Although there are some similarities between the structures of saxitoxin and of tetrodotoxin, there are also many differences.

II. CELLULAR ACTIONS

When tetrodotoxin is given to an anesthetized experimental animal, some effects are produced very rapidly. Especially striking are the precipitous fall in the arterial blood pressure and the weakening of the voluntary muscles. These changes are analogous to clinical symptomatology in cases of human poisoning. Similar changes occur with saxitoxin.

The widely distributed derangements of physiological functions in tetrodotoxin and saxitoxin poisoning can be largely attributed to their blocking actions on various types of nerve fibers. For example, on a desheathed sciatic nerve of the frog, the compound action potential is readily and reversibly blocked by tetrodotoxin or saxitoxin without any significant alterations in the resting potential (Kao and Fuhrman, 1963; Kao and Nishiyama, 1965). Although these effects are qualitatively similar to those produced by some common local anesthetic agents, such as cocaine and procaine, tetrodotoxin and saxitoxin are about 160,000 times more potent.

There is much evidence now that the block of the propagated action potential is due to an interference with the conductance change that underlies the generation of the action potential. In constant current studies on the frog sartorius muscle fiber, tetrodotoxin (Narahashi et al., 1960) and saxitoxin (Kao and Nishiyama, 1965) have both been shown to affect only the spike process without affecting the current voltage relations indicative primarily of the potassium and chloride conductances. In voltage-clamp studies on a variety of preparations, it has been shown that both toxins affect only the early transient conductance increase, which is normally associated with the inward movement of Na^+ (Narahashi et al., 1964; Takata et al., 1966; Hille, 1968; see also review of Moore and Narahashi, 1967). The conclusion is based not only on the reduction of the early inward current, but also on the absence of any effects on the kinetics of the turn-on and turn-off processes of the conductance increase (Takata et al., 1966; Hille, 1968). The toxins reduce the maximum conductance ($\bar{g}_{Na}$) and are generally believed to do so by reducing the number of membrane channels through which Na^+ can move inward (Hille, 1968). In this specificity for the Na^+ channels, the toxins differ fundamentally from the local anesthetic agents, cocaine and procaine, which also interfere with the late conductance increase normally associated with movements of K^+.

The interference with the Na^+ channel by the toxins is believed to occur in a stoichiometric relation with one toxin molecule interfering with one channel (Hille, 1968; Cuervo and Adelman, 1970). Based on this conclusion, the number of sodium channels in several excitable membranes has been estimated; in the walking leg nerve of the lobster, there are 13 channels /μm^2 of membrane surface (Moore et al., 1967); in the rabbit cervical vagus nerve, there are 75 channels/μm^2 (Keynes et al., 1971).

Although tetrodotoxin and saxitoxin share the unique specificity and potency for the Na^+ channel, they also differ in their actions on these channels. The dfference is somewhat more subtle and can be stated as follows: The dose-response relation of tetrodotoxin is rather steep. If a nerve is blocked with 4-5 times the minimum effective concentration of tetrodotoxin, it often requires quite some minutes for a reversal of the block. If even higher concentrations are used, then full recovery is often not possible. With saxitoxin, however, such poor reversals or prolonged partial block are rare; the blocks can usually be readily reversed on removal of the saxitoxin (see Kao, 1966, p.1041; Hille, 1968). Moreover, on recovery from a saxitoxin blockade, there is a tendency for the maximum sodium conductance ($\bar{g}_{Na}$) to increase beyond the original level (Hille, 1968), a phenomenon that may be associated with a transient increase in the amplitude of the action potential (Dettbarn et al., 1960; Kao and Nishiyama, 1965).

Another interesting difference between the cellular actions of tetrodotoxin and saxitoxin is found in the mammalian skeletal muscle fibers. The spike in the mammalian skeletal muscle is a Na^+ spike, which is readily blocked on removal of external Na^+ and by low concentrations of tetrodotoxin. After such a muscle is denervated, the spike is still dependent on external Na^+, but is no longer susceptible to the actions of tetrodotoxin, even at rather high concentrations of tetrodotoxin (Redfern and Thesleff, 1971). However, such tetrodotoxin-resistant spikes in the denervated muscles can be blocked by saxitoxin (Harris and Thesleff, 1971). These observations recall some similar observations on the taricha nerve and the tetrodon nerve (Kao and Fuhrman, 1967). These experiments were carried out at a time before tarichatoxin was identified with tetrodotoxin, and the experiments provided a biological proof for the identity of the two toxins (Buchwald et al., 1964). The taricha nerve was found to be about 30,000 times more resistant to tetrodotoxin, and the tetrodon nerve some 1,000 times more resistant than was the frog nerve. Yet, both nerves were about equally susceptible to saxitoxin as the frog nerve on which tetrodotoxin and saxitoxin were about equally potent. The observations on the taricha and tetrodon nerves were interpreted to mean that the membrane groups of these two preparations were sufficiently different from those of the frog nerve to make them unsusceptible to tetrodotoxin, but not so vastly different as not to react with saxitoxin. A similar interpretation has been made for the change in the reactivity to tetrodotoxin after denervation in the mammalian skeletal muscle fiber (Harris and Thesleff, 1971). Perhaps these differences are related to the difference in the reversibility of tetrodotoxin and saxitoxin blockades; and perhaps all these differences are due to a looser bonding of the saxitoxin molecule to the responsible reactive membrane groups than is in the case of the tetrodotoxin molecule. In other words, it may be asked whether these differences may not be a clue to some of the membrane groups that control the gating of sodium channels. It is probably too early to speculate further on specific details, but the potentials of these toxins as experimental tools are indeed promising.

III. SYSTEMIC ACTIONS

Tetrodotoxin and saxitoxin are both highly potent hypotensive agents, but in low doses (less than 1 μg/kg body weight) saxitoxin can sometimes cause muscular weakness without any appreciable hypotensive effect (see Kao and Nishiyama, 1965). At one time, the hypotension had been attributed to some preferential depressant actions of tetrodotoxin and saxitoxin on the medullary vasomotor center (for summary, see Kao, 1966). More recent head-body cross-perfusion experiments revealed that a medullary depressant effect was of little consequence in the hypotension, because the hypotension was produced only when the toxins were administered to the body and not when the toxins were given to the head (Kao et al., 1967). Since both toxins are known not to affect the myocardium except in rather high concentrations, the hypotension has usually been attributed to peripheral vasodilation. The question remaining is how the toxins produce the peripheral vasodilation. It should be recalled that both tetrodotoxin and saxitoxin are highly potent axonal blocking agents, and it had been suggested that a blockade of the vasomotor nerves could cause a generalized vasodilation (Kao and Fuhrman, 1963). It is now clear that in addition to a release of vasomotor tone, both tetrodotoxin and saxitoxin also cause a direct relaxant effect on the vascular smooth muscle. These effects are shown in some recent experiments using regional perfusion techniques (Lipsius et al., 1968; Kao et al., 1971), and the results can be summarized as follows: In both dogs and cats, when an innervated gracilis muscle or an innervated hind leg was perfused at a constant rate, small doses (less than 2 μg/kg) of tetrodotoxin or saxitoxin given to the body caused systemic hypotension simultaneously with a reflex vasoconstriction in the perfused musculature. If the perfusate was taken from a donor animal, unpoisoned with either toxin, the reflex vasoconstriction was particularly easy to demonstrate. Such a reflex vasoconstriction clearly could not be expected if the systemic hypotension was due to a blockade of vasomotor nerves, a process that would lead to vasodilation. When the dose of either toxin was increased (2 μg/kg or higher), then the blockade of vasomotor nerves became evident. Such a release of vasomotor tone occurred as a vasodilation that was easily distinguishable from a vasodilation attributable to a direct relaxant action of the toxins on the vasculature. The direct relaxant effect can be produced even in preparations in which all cholinergic and adrenergic responsiveness had been effectively blocked. Therefore, both tetrodotoxin and saxitoxin act similarly on the muscular vasculature directly at low doses to cause vasodilation and, additionally, at high doses on vasomotor nerves to cause a release of vasomotor tone.

Nevertheless, the systemic hypotension produced by saxitoxin differs from that produced by tetrodotoxin in that it is less severe, of shorter duration, and more often followed by a secondary pressor phase (Nagasawa et al., 1971). These differences may account for the observations that small doses of saxitoxin could cause muscular weakness without appreciable hypotensive effect (Kao and Nishiyama, 1965) and the observation that in clinical cases of paralytic shellfish poisoning, paralysis is not accompanied by recorded hypotension (see Kao, 1966, p.1033). The secondary pressor phase is apparently caused by a release of catecholamines, because it can be prevented by prior acute adrenalectomy, hexamethonium, or overnight treatment with reserpine. Whether the catecholamine is released entirely in a reflex response of the hypotension is not clear. That this reflex mechanism is involved in some instances is consistent with the observations in the regional perfusion experiments. The fact that the secondary pressor effect is more frequently produced by saxitoxin, which caused less of a pressure fall in its hypotensive phase, may be indicative of some additional but unrecognized effect of saxitoxin on the blood pressure system. Conceivably, even this difference between saxitoxin and tetrodotoxin is related to the differences they exert on the individual axons and that all these differences are manifestations of the different ways in which these two toxins react with similar membrane groups.

REFERENCES

Buchwald, H.D., Durham, L., Fischer, H.G., Harada, R., Mosher, H.S., Kao, C.Y., Fuhrman, F.A., Science 143, 474 (1964).

Cuervo, L.A., and Adelman, W.J., J. Gen. Physiol. 55, 309 (1970).

Dettbarn, W.D., Higman, H.B., Rosenberg, P., and Nachmansohn, D., Science 132, 300 (1960).

Harris, J.B., and Thesleff, S., Acta Physiol. Scand. 83, 382 (1971).

Hille, B., J. Gen. Physiol. 51, 199 (1968).

Kao, C.Y., Pharmacol. Rev. 18, 997 (1966).

Kao, C.Y., Fed. Proc. (in press).

Kao, C.Y., and Fuhrman, F.A., J. Pharmacol. Exp. Ther. 140, 31 (1963).

Kao, C.Y., and Fuhrman, F.A., Toxicon 5, 25 (1967).

Kao, C.Y., and Nishiyama, A., J. Physiol. (London) 180, 50 (1965).

Kao, C.Y., Nagasawa, J., Spiegelstein, M.Y., and Cha, Y. N., J. Pharmacol. Exp. Ther. 178, 110 (1971).

Kao, C.Y., Suzuki, T., Kleinhaus, A.L., and Siegman, M.J., Arch. Int. Pharmacodyn. Therap. 165, 438 (1967).

Keynes, R.D., Ritchie, J.M., and Rojas, E., J. Physiol. (London) 213, 235 (1971).

Lipsius, M.R., Siegman, M.J., and Kao, C.Y., J. Pharmacol. Exp. Ther. 164, 60 (1968).

Moore, J.W., and Narahashi, T., Fed. Proc. Fed. Amer. Soc. Exp. Biol. 26, 1655 (1967).

Moore, J.W., Narahashi, T., and Shaw, T.I., J. Physiol. (London) 188, 99 (1967).

Mosher, H.S., Fuhrman, F.A., Buchwald, H.D., and Fischer, H.G. Science 144, 1100 (1964).

Nagasawa, J., Spiegelstein, M.Y., and Kao, C.Y., J. Pharmacol. Exp. Ther. 178, 103 (1971).

Narahashi, T., Deguchi, T., Urakawa, N., and Ohkubo, Y., Amer. J. Physiol. 198, 934 (1960).

Narahashi, T., Moore, J.W., and Scott, W., J. Gen. Physiol. 47, 965 (1964).

Redfern, P., and Thesleff, S., Acta Physiol. Scand. 82, 70 (1971).

Schantz, E.J., Ann. N.Y. Acad. Sci. 90, 843 (1960).

Schantz, E.J., Lynch, J.M., Vayada, G., Matsumoto, K., and Rapoport, H., Biochem. 5, 1191 (1966).

Takata, M., Moore, J.W., Kao, C.Y., and Fuhrman, F.A., J. Gen. Physiol. 49, 967 (1966).

Wong, J.L., Oesterling, R., and Rapoport, H., J. Amer. Chem. Soc. 93, 7344 (1971).

Chapter 6

NEMATOCYST TOXINS OF COELENTERATES

Charles E. Lane

Rosenstiel School of Marine and Atmospheric Science
University of Miami
Miami, Florida

The nematocyst is a characteristic intracellular organoid of coelenterates; when these structures are found in other animals, as in the cerata of Glaucus marinus and some other nudibranchs (Thompson and Bennett, 1969) or on the dorsal arms of the octopus, Tremoctopus violaceus (Jones, 1963) they may always be traced to a coelenterate origin. Nematocysts are salvaged by some marine flatworms and used as defensive weapons.

Nematocysts occur in specialized epithelial cells called cnidoblasts and always appear to be produced from Golgi material (Slautterback and Fawcett, 1959; Chapman, 1961; Sutton and Burnett, 1969; Westfall, 1965, 1966, 1970a). They consist, essentially, of a capsule wall enclosing a coiled hollow tubule. Other components of the nematocyst complex may include a cap-like operculum and a cnidocil with an internal organization resembling that of cilia, thought to function like a trigger. Nematocysts range in size from only a few microns in diameter to those of Halistemma rubrum that measure 1.12 x 0.12 mm (Iwanzoff, 1896), making them the largest known. The thick capsule wall appears to thin abruptly and turn inward at one pole to form the hollow nematocyst tubule. The capsule also encloses a fluid rich in solutes, the staining reaction of which suggests that it contains considerable dissolved protein (Boisseau, 1948, 1952). The fluid contents of the nematocysts of Corynactis are rich in osmotically active solutes that depress the freezing point to -6°C. (Picken and Skaer, 1966).

The characteristic protein toxin that arms the nematocyst and augments its effectiveness is dissolved in this solution.

Evidence continues to mount suggesting that nematocysts may not be the completely independent effectors that earlier investigators (Phisalix, 1922; Parker and Van Alstyne, 1932, among others) thought them to be. Westfall (1970a, b) has published convincing electron micrographs showing synapselike structures with well-developed synaptic vesicles proximal to the presynaptic membrane in a variety of coelenterates. Each class of the phylum is represented in her study, and in representatives of each class, Westfall has shown these characteristic structures. Interneuronal synapses as well as neuron-cnidoblast synapses are shown. Electrophysiological evidence positively identifying these structures with impulse transmission is still lacking. Electron microscopy does not permit differentiating between afferent and efferent pathways in primitive nervous systems, hence no conclusion about the nature or location of receptor endings is possible at this time. The morphological basis for neural function, however, has been clearly established. Studies of Ross and his colleagues (Davenport et al., 1961; Ross and Sutton, 1961; Ross, 1965) strongly suggested that nematocysts of anemones Calliactis parasitica and Stomphia coccinea discharged preferentially against certain substrates; Calliactis on shells of the hermit crab Pagurus, and Stomphia on the shells of the mussel Modiolus. Ross and Sutton (1961) identified an insoluble highly stable organic component of the shell and periostracum of the mollusc as the primary stimulus for nematocyst discharge. These observations suggest that these cnidarians have specialized receptors and essential transmitter mechanisms mediating these responses. Westfall's results appear to provide a structural framework for this function.

Nematocysts discharge readily in response to many physical and chemical stimuli when they are in their normal location within the cnidoblast and when, presumably, all parts of the nematocyst complex are intact and functional. Accessory structures within the cell, apart from possible innervation that may influence discharge, include a basketlike network of fibers that encircle the intact capsule, the cnidocil, and the operculum. The precise role these structures may play in discharge has not yet been clarified. Several observations suggest that the nematocyst is more resistant to discharge after it has been separated from the cnidoblast (Burnett et al., 1968; Yanagita, 1969). This characteristic may account for their failure to discharge in passage through the digestive tract of flatworms or through the complicated pathway they must follow to arrive at the tip of the cerata of Glaucus. In the Man-of-War fish Nomeus gronovii, Mr. Charles Mayo in my laboratory has observed masses of undischarged nematocysts in the small intestine. These have clearly been freed from their nematocytes during digestion of tentacle fragments ripped from the host. The relationship between Physalia physalis and Nomeus gronovii clearly does not fit the classic concept of symbiosis, a mutually beneficial relationship, because in this association the fish is an obvious predator.

When activated, by whatever mechanism, the nematocyst discharges with explosive violence. During discharge the coiled internal tubule everts progressively, bringing its armament of chitinous spikes and spines to the definitive external surface. The surface ornamentation serves as an effective tangle to ensnare small crustacean prey organisms. Actual penetration of the prey is facilitated by the continuously renewed crest of spines created at the tip of the everting tubule when spines that were previously internal reach the tip and become superficial. Robson (1953) described this process in these terms: ''The tip of the shaft is formed by a constantly renewed spearhead of opposed barbs. These flick out sharply and take their positions in each of the spiral rows.'' In the Portuguese Man-of-War, Physalia physalis, eversion of the nematocyst tubule is sufficiently forceful to penetrate a fish scale, the chitinous exoskeleton of crustacean prey, or the skin of a human investigator even when he is protected by a surgical glove. The extruded tubule of a P. physalis nematocyst 20 μm in diameter may exceed 3000 μm in length (Lane and Dodge, 1958); that of the cubomedusan Chironex fleckeri of the Indo-Pacific penetrates all layers of the human skin (Cleland and Southcott, 1965) delivering its toxic charge into the richly vascularized dermis.

Although the stinging powers of anemones and some jelly fish have been well known to fishermen since the time of Aristotle, the nematocysts responsible for these powers were apparently first seen in the freshwater coelenterate Hydra by Trembley (1744). The stinging powers of the Portuguese Man-of-War, Physalia, have been well known to mariners for some centuries, at least. In 1579, the Jesuit Thomas Stevens, describing his voyage around the coast of Africa, included the following comment in a letter to his father (in Hakluyt's Voyages, 1904, Vol.6, p.379): ''Along all that coast we oftentimes saw a thing swimming on the water like a cock's comb, which they call a Ship of Guinea, but the colour much fairer. Which comb standeth upon a thing like the swimmer of a fish in colour and bigness, and beareth underneath in the water strings which save it from turning over. This thing is so poisonous that a man cannot touch it without great peril.''

The stinging capability of the Milleporina was graphically described by Charles Darwin in The Voyage of the Beagle (1845). In April 1836 the Beagle was at Keeling Island. Darwin writes (p.464): ''I was a good deal surprised by finding two species of coral of the genus Millepora (M. complanta and M. alcicornis) possessed of the power of stinging... The stinging property seems to vary in different specimens; when a piece was pressed or rubbed on the tender skin of the face or arm, a prickling sensation was usually caused, which came on after the interval of a second, and lasted only for a few minutes. One day, however, by merely touching my face with one of the branches, pain was instantaneously caused; it increased as usual for a few seconds and, remaining sharp for some minutes, was perceptible for half an hour afterwards. The sensation was as bad as that from a nettle, but more like that caused by the Physalia or Portuguese Man-of-war. Little red spots were produced on the tender skin of the arm, which appeared as if they would have formed watery pustules, but did not.''

Even though nematocysts were recognized as the stinging elements of coelenterates more than 200 years ago, serious study of this capability has been limited to the present century.

Early studies of coelenterate toxicity employed very crude extracts of whole animals, entire tentacles (Richet et al., 1902; Richet, 1903; Shapiro, 1968a; Keen and Crone, 1969), or entire mesenteries or acontia (Welsh, 1955). Toxic symptoms evoked by injecting such extracts into experimental animals have been attributed to compounds variously called "hypnotoxine," "congestine," and "thalassine" (Richet, 1903; Cantacuzene and Damboviceanu, 1934). The term "anaphylaxis" was coined by Portier and Richet (1902, 1936) in their early studies of coelenterate toxicity to describe the sensitivity induced by a sublethal dose of toxin. More recently, tetramine, trigonelline, and homarine have been identified in crude homogenates of entire Actinia equina and of Anemonia sulcata (Ackerman et al., 1923, Ackerman, 1953). Welsh (1955) has identified urocanylcholine and other quarternary amines in crude extracts of Condylactis gigantea, Metridium dianthus, and Cyanea capillata. Shapiro (1968a) described the preparation and purification of a single toxic protein (mol wt about 12,000) from tentacles of Condylactis gigantea. Toxin solutions were made from "redissolved acetone powder" consisting of all the material in the tentacle homogenate insoluble in 80% acetone. Martin (1960, 1967) described a heat-sensitive, nondialyzable, antigenic, curarelike principle in the anemone Rhodactis howesii, to which he ascribed its considerable human toxicity. This observation acquires added significance since Rhodactis is commonly used for food in the Indo-Pacific region even though it may cause death when ingested raw. Cooking in water however extracts the toxic components, rendering it innocuous. It should be emphasized that all these studies used relatively nonspecific extracts of complex organs or tissues. In none of the studies thus far described could the biological activity of the extracts be ascribed exclusively to the nematocysts.

Modern-day studies of coelenterate toxins originated with Phillips' description (1956) of a method for isolating clean nematocysts of Metridium senile from the tissues in which they occur. His method and various modifications of it (Lane and Dodge, 1958; Burnett et al., 1968; Endean et al., 1969; Goldner et al., 1969) employ proteolytic enzymes to liberate nematocysts from the cells that contain them. The undischarged nematocysts, being heavier than the rest of the tissue components in the hydrolysate, settle into a compact layer. Repeated washing of the suspension removes the last fragments of tentacle tissue and all soluble extranematocyst compounds. Toxin loss during this process is minimal. The concentrated frozen nematocysts may be stored for several months without significant decrease in the activity of the contained toxin. Homogenization or sonication of the frozen nematocysts fragments the capsule wall, permitting the fluid toxin to escape into the medium. Centrifugation separates capsule and tubule fragments from the toxic supernatant solution. Some loss of activity occurs when the toxin solution is lyophilized, but the residual activity in the white fluffy powder persists for several months at $-20^{o}C$.

Most of the pharmacological studies to be described in this survey have been made on extracts prepared in this way. An interesting technique that provides only authentic nematocyst toxin has been described by Barnes (1966, 1967) to sample the particularly virulent toxin of the sea wasp, Chironex fleckeri. Living tentacles are draped over a stretched piece of human amnion and stimulated electrically. Discharging nematocysts deliver their toxin through the entire thickness of the amnion, from the under surface of which it may be recovered.

Naturally this procedure varies in effectiveness with the activity and morphology of the nematocysts studied. Nematocysts of Aurelia aurita, for example, rarely evoke local reactions in man; in our experience, those of Velella and Porpita never do. Nematocysts of these animals could not be studied by Barnes' procedure, only because they could not penetrate the thickness of the amnion.

Early workers (see Hyman, 1940) assumed that the wall of the nematocyst capsule was chitinous. Lenhoff et al. (1957) detected hydroxyproline among the amino acids in hydrolysates of discharged nematocysts of Hydra. Lenhoff and Kline (1958) identified proline, the other imino acid, in capsule hydrolysates. Since large amounts of hydroxyproline and of proline are characteristic of vertebrate collagen these authors suggested that the capsule wall belonged to this class of compounds. Phillips (1956) had earlier suggested that Metridium nematocyst capsules contained cartilaginous material. Hydra nematocysts (Kline, 1961) are birefringent; resistant to tryptic hydrolysis; and heat stable, retaining their characteristic morphology even after "many hours in the autoclave at 121°."

The amino acid composition of Physalia nematocysts, freed from tentacular tissue, ruptured by mechanical homogenization and washed exhaustively by centrifugation, is shown in Table I (from Lane, 1968). It is clear that Physalia nematocysts, similar to those of Hydra, contain large amounts of imino acids. The capsules of Physalia contain approximately 30% ash on a dry basis (Table II). In addition to the common elements one would expect to find in a pelagic marine organism the very high content of boron is noteworthy. Iron, aluminum, zinc, and copper

TABLE I

Amino Acids in Physalia Nematocysts (μmole/mg Dry Matter)

	ASP	THR	SER	GA	PRO	GLY	ALA	VAL
Toxin	0.104	0.049	0.072	0.715	0.189	0.161	0.156	0.070
Capsule	1.033	0.876	0.491	1.069	3.195	7.445	5.379	1.065

	MET	ILEU	LEU	TYR	PHE	LYS	HIS	ARG
Toxin	0.016	0.056	0.071	0.031	0.039	0.040	0.120	0.027
Capsule	0.631	0.719	0.559	0.199	0.228	0.377	0.467	0.154

TABLE II

Composition of Physalia physalis (ppm of Ash)

Element	Toxin	Washed nematocyst capsule
Ag	<3	50
Ba	100	90
Co	<5	20
Cr	<10	40
Cu	60	900
Mo	<5	10
Ni	22	950
Pb	<20	90
Sr	800	1,500
Ti	6	130
V	<5	10
Y	<3	<3
Zn	<100	1,100
B	1,225	10,780
CaO	11,000	210,000
SiO_2	10,200	530,000
MgO	120,000	20,000
Al_2O_3	500	10,300
Fe	157	1,362
Mn	12	1,100

contribute significantly to the total inorganic composition. Silicon, calcium, and magnesium are also abundant. The capsule wall is negative for Benedict, Molisch, and other qualitative carbohydrate tests and appears to be lipid free. Nematocysts of Chrysaora quinquecirrha, on the contrary, were reported to contain 3.74% hexosamine and 1.87% reducing sugars, as well as 2.09% cholesterol (Goldner et al., 1969). Since this analytical sample also contained 27% nucleic acids, it must have been grossly contaminated with cellular debris from the tentacle.

Washed, broken, empty nematocysts of other cnidarians have apparently not been analyzed. It is to be hoped that definitive analyses of both Chrysaora and Chironex nematocysts may shortly be available.

Our studies (Lane and Dodge, 1958; Lane et al., 1961; Lane and Larson, 1965; Larson and Lane, 1966; Lane, 1967a,b; Stillway and Lane, 1971) have established that the nematocyst toxin of Physalia is a complex protein consisting of several peptides and that it is thermolabile and nondialyzable. The toxin includes phospholipase and at least one protease active above pH 7..0 with substrate specificity different from trypsin. Analysis of various preparations has yielded nitrogen values ranging from 14 to 16% after correction for inorganic ash. The toxin is negative for all qualitative carbyhydrate tests but is strongly ninhydrin and biuret positive. The amino acid composition of the toxin (Table I) is unusual only in the high concentration of glutamic acid and proline.

For *Chironex fleckeri*, Endean et al. (1969) present histochemical evidence suggesting that nematocysts contain proteins, 3-indoyl derivatives, sulfur amino acids, carbohydrate-containing proteins, and metachromatic substances. No lipids were detected. These data suggest that the capsule contents of *Chironex* may be qualitatively similar to those of *Physalia*.

When nematocysts of *C. fleckeri* were isolated by mincing thawed tentacle in 7% sucrose, the suspending medium approached the toxicity of an homogenate of the isolated capsules (Freeman and Turner, 1969), suggesting either that the capsule wall was permeable to the toxin or that many nematocysts discharged into the salt-free suspending medium. *Physalia* nematocysts in contrast may be washed exhaustively in seawater without apparent loss of activity (Lane and Dodge, 1958) and without inducing significant discharge.

The nematocyst toxin of *C. fleckeri* has been studied recently by several different investigators (Baxter et al., 1968; Freeman and Turner, 1969; Endean et al., 1969; Crone and Keen, 1969, 1971; Baxter and Marr, 1969; Endean and Henderson, 1969). This toxin appears to be a complex, causing hemolysis as well as cardiovascular, dermonecrotic, and membrane permeability changes. Most investigators have described cardiac arrhythmia in treated animals ranging in severity from augmentation of the R peak to total atrioventricular blockade. Endean and his co-workers describe "ventricular contracture" as one of the terminal effects of envenomation, while Freeman and Turner observed that the mouse heart was still beating at autopsy, but with a 3.1 AV block. Freeman and Turner, and Baxter and Marr document the hemolytic properties of *C. fleckeri* venom for the mouse, rabbit, cat, monkey, sheep, dog, man, goat, echidna, horse, trout, chicken, and pigeon. Erythrocytes of various species are lysed by different venom concentrations, but the phenomenon of hemolysis is firmly established. Mouse red blood cells require twice the venom concentration for lysis as human erythrocytes, four times as much as the goat, and more than 150 times as much as horse red blood cells.

Since the hemolysin was not activated by incubation with egg yolk, it does not resemble a phospholipase. Baxter and Marr exposed susceptible erythrocytes to blood hemolyzed by *Chironex* toxin and observed no further hemolysis. They concluded that the hemolysin is adsorbed on the erythrocyte membrane causing lysis, the hemolytic agent being inactivated in the process. Molecular weight studies with Sephadex and the ultracentrifuge suggest that the biologically active molecules in *C. fleckeri* toxin have a mol wt between 10,000 and 30,000. The toxin is thermolabile, being inactivated at 42°C.

Intravenous injection of *C. fleckeri* toxin increased serum K^+ levels in the rat without significant alteration in serum Na^+. Subcutaneous, intramuscular, or intraperitoneal injection was very much less effective than intravenous administration. There was no evidence that histamine was released from the skin after intradermal injection of up to five intravenous lethal doses in the rat and the mouse. Neuromuscular transmission was not affected by *C. fleckeri* toxin and the respiratory center in experimental animals remained competent until death. General effects of *C. fleckeri* toxin could result from alteration in membrane permeability and active transport capability. Endean and Henderson attribute the terminal contracture of mammalian ventricular muscle, of the depressor muscles of the barnacle *Balanus nigrescens*, and of the rat uterus to increased permeability of the muscle membranes, permitting efflux of intracellular Ca^{2+}. Paralysis in their view results when the intracellular Ca^{2+} level is depressed below a limiting value.

Baxter et al. (1968) described the antigenic property of *C. fleckeri* toxin. Neutralizing antibody was produced by both rabbits and mice after repeated injection of sub-lethal doses of toxin. This treatment immunized experimental animals against the lethal component(s) of the toxin but afforded no protection against the necrotic reaction. These results suggest that an effective antivenin may be developed to treat human victims of *C. fleckeri* stings. As Baxter et al. conclude, however, in cases of severe stings the victim dies so quickly that there is no time for effective treatment. Clinical descriptions of human fatalities from *C. fleckeri* stings (Cleland and Southcott, 1965; Barnes, 1966) emphasize how suddenly the victim may die. Barnes says: "From the moment of stings to apparent death, the interval may be but a few minutes. Three minutes is a common assessment, but lesser periods are well authenticated..." The victims show no generalized paralysis, no respiratory failure, and no cyanosis. Rather they are gray, apprehensive, and conscious until the final moment of collapse.

Most studies of C. fleckeri toxin have been marred by imprecision in dose level determination. Endean expresses dose levels on the basis of the number of nematocysts homogenized to yield the preparation used. Other investigators have determined the volume of extract required to kill half the mice injected intravenously after a fixed time interval and then expressed dosage in terms of numbers of mouse lethal doses. Both of these systems are cumbersome and could readily be improved by expressing dosage in terms of total dry weight, protein, or nitrogen, then C. fleckeri data might be more readily compared with those from other cnidaria.

Shapiro (1968 a,b) obtained a single toxic protein (mol wt about 12,000) from Condylactis gigantea by redissolving an acetone powder of a tentacle homogenate. This toxin produced an immediate rigid paralysis in crustacea that was followed by a flaccid phase. The paralysis results from toxin-induced rhythmic firing in all axons studied, taking the form either of high frequency bursts lasting several seconds or low frequency trains of short duration (Shapiro and Lilleheil, 1969a). Further studies showed that the toxin transforms normal crustacean action potentials into prolonged plateau potentials lasting several seconds (Shapiro and Lilleheil, 1969b). They attribute the plateau phenomenon to greatly increased sodium permeability of the poisoned axons (Lilleheil and Shapiro, 1969).

The nematocyst toxin of Physalia physalis, the Portuguese Man-of-War, is a thermolabile nondialyzable complex protein containing several peptides (Lane and Dodge, 1958; Lane, 1967a; Lane, 1961; Hastings et al., 1967). It is lethal to all metazoan animals tested, but is an acceptable carbon source for some protozoa, promoting the growth of Paramecium and Tetrahymena (Lane et al., 1961). The intravenous LD_{50} for the dog is approximately 200 μg/kg of lyophilized toxin, for the rat 100 μg/kg, and for the white mouse approximately 50 μg/kg (Lane, 1967a).

Both the neurogenic heart of crustacea (Lane and Larsen, 1965) and the myogenic heart of the rat (Larsen and Lane, 1966) and the dog (Hastings et al., 1967) show dose-dependent damage. Less than 60 sec after intravenous injection of as little toxin as 12 μg/kg in the dog, the P–R interval in the electrocardiogram was reduced; then the P wave was suppressed, with the activation of an ectopic pacemaker site in the vicinity of the A–V node. This ectopic center dominated the heart for variable periods up to 1 min, after which normal sinus rhythm was restored spontaneously. Larger doses of toxin (50–100 μg/kg) elicited coupled extrasystoles similar to those induced by ouabain. These abnormalities were corrected and normal sinus rhythm restored by the intravenous administration of isotonic KCl solution. Toxin at 200 μg/kg quickly caused cardiovascular collapse.

The rat shows a series of similar effects after toxin injection, including prolongation of the Q–T intervals at medium dose levels (100–300 μg/kg). Higher dose levels suppressed the P wave, caused A–V block, ventricular insufficiency, fibrillation, and death (Larsen and Lane, 1966). This entire spectrum of changes may be completed in less than 10 sec in the rat when the dose level is 100 μg/kg or higher.

The neurogenic heart of crustacea responds somewhat differently to *Physalia* toxin (Lane and Larsen, 1965). In the land crab, *Cardisoma guanhumi*, the heart is immediately arrested following toxin injection. The output of bioelectric potentials by the neurons of the cardiac ganglion is exaggerated, but uncoordinated, and without effect on the contractile mechanisms of the heart. After the injection of sublethal doses of toxin into the hemocoele, effects on the cardiac ganglia may often be detected for an hour or more. At higher dose levels the cardiac arrest is irreversible. If the arrested heart be driven by stimuli applied directly through indwelling electrodes, it responds with a normal beat, suggesting that muscle cells are competent but that conduction along the axons of the cardiac ganglion cells has been prevented or that the terminations of these processes on the heart muscle cells have been altered.

In rats treated with various sublethal doses of *Physalia* nematocyst toxin, all blood samples were massively hemolyzed. In treated dogs (Hastings et al., 1967), all blood samples were hemolyzed and serum K^+ levels were significantly elevated. Serum Na^+ values tended to be lower than control values. Changes in serum electrolytes could not be entirely accounted for by red blood cell destruction.

When *Physalia* toxin was applied to a segment of frog sciatic nerve in concentrations greater than 1.0 mg/ml of Ringer's solution, conduction through the treated segment was blocked (Larsen and Lane, 1970a). Stimulation of the nerve distal to the treated segment caused normal contraction of the attached gastrocnemius muscle. After a segment of the nerve had been desheathed and then simularly treated, the minimal effective dose of toxin was 0.25 mg/ml. Gastrocnemius preparations, with all neuromuscular connections blocked by curare, failed to contract in response to direct stimulation after about 10 min of exposure to 1.0 mg/ml of toxin.

The effects of P. physalis toxin on active Na^+ transport and membrane permeability were investigated using isolated short-circuited preparations of frog skin (Larsen and Lane, 1970b). When toxin was added to the solution bathing the inner surface, it caused an increase in the rate of Na^+ transport across the skin. It was without effect when added to the solution bathing the outer surface of the skin. Ouabain inhibited the toxin effect. The toxin reversed the inhibition of the Na^+ transport normally caused by high concentrations of Ca^{2+}.

Unpublished observations of Dr. D.J. Quinn in the author's laboratory suggest that P. physalis toxin, like ouabain, may inhibit ATPases in the gill epithelium of the land crab, Cardisoma guanhumi, thus reducing active ion transport and preventing ion regulation when the experimental animals are subjected to ionic challenge.

Physalia toxin added to the medium at 20 μg/ml caused prompt contraction of segments of rat or guinea pig ileum (Lane, 1967a) similar in magnitude to that following administration of 0.05 μg/ml acetylcholine. However, neither incubation of toxin with active acetylcholinesterase nor treatment of the preparation with atropine modified the characteristic response to the toxin. Histamine is another agent that causes contraction of isolated ileal preparations, but the addition of an antihistamine to the bath did not modify the toxin-induced contraction of the gut. Serotonin, which also causes contraction of intestinal preparations, was shown by chromatography to constitute less than 0.05% of the toxin. Hexamethonium, in a dose that blocked the ganglionic stimulating property of nicotine, did not affect the toxin-induced contraction. From these observations we concluded (Garriott and Lane, 1969) that the stimulating action of Physalin toxin on the isolated intestine is not because of the presence of histamine, acetylcholine, or serotonin and is probably not ganglionic in origin.

These various observations suggest that Physalia toxin owes its general biological effectiveness to its ability to modify membrane transport phenomena.

REFERENCES

Ackerman, D., Hoppe-Seyl. Z. 295, 1 (1953).

Ackerman, D., Holtz, F., and Reinwein, H., Z. Biol. 79, 113 (1923).

Barnes, J.H., In "The Cnidaria and their Evolution," Symp. Zool. Soc., London, Vol. 16, pp. 307-332. Academic Press, New York, 1966.

Barnes, J.H., In "Animal Toxins" (Findlay E. Russell and Paul R. Saunders, eds.), pp. 115-131. Pergamon Press, London 1967.

Baxter, E.H., Marr, H.G., and Lane, W.R., Toxicon 6, 45 (1968).

Baxter, E.H., and Marr, A.G.M., Toxicon 7, 195 (1969).

Boisseau, J.P., Compt. Rend. 13^{e} Congr. Int. Zool., Paris (1948).

Boisseau, J.P., Bull. Soc. Zool. Fr. 77, 151 (1952).

Burnett, J.W., Stone, J.H., Pierce, L.H., Cargo, D.G., Layne, E.C., and Sutton, J.S., J. Invest. Derm. 51, 330 (1968).

Cantacuzene, J., and Damboviceanu, A., Compt. Rend. Soc. Biol., Paris. 117, 138 (1934).

Chapman, G.B., In "The Biology of Hydra" (H.M. Lenhoff and W.F. Loomis, eds.) pp. 131-151 (1961).

Cleland, J.B., and Southcott, R.V., "Injuries to Man from Marine Invertebrates in the Australian Region," Commonwealth of Australia, Canberra. University of Miami Press, 1965.

Crone, H.D., and Keen, T.E.B., Toxicon 7, 79 (1969).

Crone, H.D., and Keen, T.E.B., Toxicon 9, 145 (1971).

Darwin, C. "Journal of Researches into the Natural History and Geology of the Countries visited during the Voyage of HMS Beagle around the World under the Command of Capt. Fitzroy, R.N.," 2nd ed., 1845. John Murray, London (1st ed. 1839).

Davenport, D., Ross, D.M., and Sutton, L., Vie et Milieu 12, (2), 197 (1961).

Endean, R., and Henderson, Lyn, Toxicon 7, 303 (1969).

Endean, R., Duchemin, C., McColm, D., and Fraser, E.H., Toxicon 6, 180 (1969).

Freeman, Shirley E., and Turner, R.J., Brit. J. Pharmac. 35, 510 (1969).

Garriott, J.C., and Lane, C.E., Toxicon 6, 281 (1969).

Goldner, R., Burnett, J.W., Stone, J.S., and Dilaimy, M.S., Proc. Soc. Exp. Biol. Med. 131, 1386 (1969).

Hakluyt, R., "The principal Navigations Voyages Traffiques & Discoveries of the English Nation, Made by Sea or Overland to the Remote and Farthest Distant Quarters of the Earth at any Time within the Compass of These 1600 Years", 12 Vols. Vol.6, pp. XVI and 1-527. Glasgow University Press, 1904.

Hastings, L.G., Larsen, J.B., and Lane, C.E., Proc. Soc. Exp. Biol. Med. 125, 41 (1967).

Hyman, L.H., The Invertebrates: Protozoa Through Ctenophora. McGraw Hill Book Co., New York, 1940.

Iwanzoff, N., Bull. Soc. Nat. Moscow, N.S. 10, 95 and 323 (1896).

Jones, E.C., Science 139, 764 (1963).

Keen, T.E.B., and Crone, H.D., Toxicon 7, 55 (1969).

Kline, E.S., In "The Biology of Hydra and Some Other Coelenterates" (H.M. Lenhoff and W.F. Loomis, eds.) Vol. 1, pp.153-166. University of Miami Press, 1961.

Lane, C.E., In "The Biology of Hydra and Some Other Coelenterates" (H.M. Lenhoff and W.F. Loomis, eds.), pp. 169-178. University of Miami Press, 1961.

Lane, C.E., In "Animal Toxins" (F.E. Russell and P.R. Saunders, eds.). Pergamon Press, New York, 1967a.

Lane, C.E., Fed. Proc. 16, 1225 (1967b).

Lane, C.E., In "Chemical Zoology" (M. Florkin and B.T. Scheer, eds.), Vol.2. Academic Press, New York, 1968.

Lane, C.E., Coursen, B.W., Hines, K., Proc. Soc. Exp. Biol. Med. 107, 670 (1961).

Lane, C.E., and Dodge, Eleanor, Biol. Bull. 115, 219 (1958).

Lane, C.E., and Larsen, J.B., Toxicon 3, 69 (1965).

Larsen, J.B., and Lane, C.E., Toxicon 4, 199 (1966).

Larsen, J.B., and Lane, C.E., Toxicon 8, 21 (1970a).

Larsen, J.B., and Lane, C.E., Comp. Biochem. Physiol. 34, 333 (1970b).

Lenhoff, H.M., and Kline, E.S., Anat. Rec. 130, 425 (1958).

Lenhoff, H.M., Kline, E.L., and Hurley, R., Biochem. Biophys. Acta.. 26, 204 (1957).

Lilleheil, G., and Shapiro, B.I., Comp. Biochem. Physiol. 30, 281 (1969).

Martin, E.J., Pacific Sci. 14, 403 (1960).

Martin, E.J., Int. Archs. Allergy Appl. Immun. 32, 342 (1967).

Parker, G.H., and Van Alstyne, M.A., J. Exp. Zool. 63, 329 (1932).

Phillips, J.H., Jr., Nature 178, 932 (1956).

Phisalix, Marie, "Animaux Venimeux et Venins." G. Masson, Paris, 1922.

Picken, L.E.R., and Skaer, R.J., In "The Cnidaria and Their Evolution." Zool. Soc. of London Symp., pp.19-50, Academic Press, New York, 1966.

Portier, P., and Richet, C., Compt. Rend. 154, 247 (1902),

Richet, C., Compt. Rend. Soc. Biol., Paris 55, 246 (1903).

Richet, C., and Portier, P., Resultats des campagnes scientifiques, Monaco 95, 3 (1936).

Richet, C., Perret, A., and Portier, P., Compt. Rend. 154, 788 (1902).
Robson, E.A., Quart. J. Microscop. Sci. 94, 229 (1953).
Ross, D.M., and Sutton, L., Proc. Roy. Soc. B 155, 266 (1961).
Ross, D.M., Science 148, 527 (1965).
Shapiro, B.I., Toxicon 5, 253 (1968a).
Shapiro, B.I., Comp. Biochem. Physiol. 27, 519 (1968b).
Shapiro, B.I.,and Lilleheil, G., Comp. Biochem. Physiol. 28, 1225 (1969a).
Shapiro, B.I., and Lilleheil, G., Comp. Biochem. Physiol. 28, 1225 (1969b).
Slautterback, D.L., and Fawcett, D.W., J. Biophys. Biochem. Cytol. 5, 441 (1959).
Stillway, L.W., and Lane, C.E., Toxicon 9, 193 (1971).
Sutton, J.S., and Burnett, J.W., J. Ultrastructure Res. 28, 214 (1969).
Thompson, T.E., and Bennett, I., Science 166, 1532 (1969).
Trembley, A., Memoires pour servir à l'histoire d'un genre de polypes d'eau douce, ā bras en forme de cornes. Leide (Verbeek) p.246. (1744).
Welsh, J.H., Deep Sea Res. Suppl. 3, 287 (1955).
Westfall, Jane A., Amer. Zoologist. 5, 377 (1965).
Westfall, Jane A., Zeitsch f. Zellforschung, 75, 381 (1966).
Westfall, Jane A., J. Ultrastructure Res. 32, 237 (1970a).
Westfall, Jane A., Z. Zellforsch. 110, 457 (1970b).
Yanagita, Tame Masa., Bull. Mar. Biol. 13, 221 (1969).

Chapter 7

SOURCE OF THE TOXICITY OF PUFFER FISHES, GENUS SPHOEROIDES

Edward Larson, Jack Grossman, John Humphries
and John Klinovsky

Department of Biology
University of Miami
Coral Gables, Florida
and
The Miami Seaquarium
Miami, Florida

I. HISTORY

The toxic principle, tetrodotoxin (TTX) and other possible toxic principles (Lalone et al., 1963), present in the flesh, skin, gonads, and liver of Sphoeroides nephelus,* could be either exogenus or endogenus. Halstead (1959) has a diagram illustrating a theory for the origin of fish poisons i.e., plant containing poison or precursor eaten by a herbivore and then the herbivore eaten by another fish, which is eaten by man. Habekost et al. (1955) found that some algal extracts (Bryopsis and Enteromorpha) were weakly toxic when injected intraperitoneally into mice. Williams (1950) mentions the use of seaweeds for animal nutrition. Tressler and Lemon (1951) report the extensive use of algae as human food in the Orient and also the utilization of algae for the production of extractives.

* The nomenclature of Robert L. Shipp and Ralph W. Yerger [Copeia No. 3, 425 (1969)] has been adopted in place of the older designation, Sphoeroides maculatus.

II. METHODS

Homogenates (proportion: 1 gm alga to 1 ml distilled water) of the three predominant algae* (Bryopsis hypnoides, Gracilaria compressa, and Ulva lactuca) from the Titusville, Florida, fishing pier area,** were prepared after the algae had been washed several times with distilled water to remove sand, salts and other possible contaminants. These homogenates were administered, 1 ml/20 gm body weight, via stomach tube to hamsters lightly anesthetized with ether.

Mice were injected intraperitoneally, 1 ml/20 gm body weight, with extracts prepared from a homogenized, centrifuged, and filtered extract of the three previously mentioned algae (proportion: 1 gm algae to 2 ml water). Before the extracts were prepared, the algae were washed as previously described. These animals were observed for 72 hr (Lalone et al., 1963). Bacteria-free extracts*** of these algal preparations also were injected in the same manner.

Three groups of young male Leghorn chicks were fed an equal mixture of 10% of their body weight of the three previously mentioned ground algae (washed as before) and Purina Startena each day for 10 consecutive days. A fourth group was fed an equal mixture of ground culinary grade leaf lettuce (Lactuca sativa) and Purina Startena as a control. These animals were weighed, rectal temperatures obtained, and respirations counted at the start, middle, and end of the experiment. The general condition and physical development of these animals also were observed during the experimental period.

The gastrointestinal tracts of 38 Sphoeroides nephelus, southern puffer (Titusville area), 87 Sphoeroides testudineus, checkered puffer (Baker's Haulover and Turkey Point, Miami area), and 12 Arius felis, sea catfish (Titusville and Miami

* Identified by Dr. Robert H. Williams, Biology Dept., Univ. of Miami, and classified according to Taylor, W.R., "Marine Algae of the Eastern Tropical and Subtropical Coasts of the Americas," p.662. University of Michigan Press, Ann Arbor, 1960.

** The present source of the toxic fish, Sphoeroides nephelus.

*** Prepared by Dr. Bennett Sallman, Chairman, Dept. of Microbiology, University of Miami.

areas) were examined* under a dissecting microscope and the mollusks found therein classified according to Abbott (1953).

Brachidontes exustus and Chione cancellata from the Miami Biscayne Bay area were made into slurries by breaking the shells and then semihomogenizing the entire mollusk for a few minutes in a Waring blender with an equal volume of distilled water. After a few hours fast, the chicks (male Leghorn) would consume 5% of their body weight of the slurry, sprinkled with a small amount of Purina Startena, within a period of an hour or two.

On three separate occasions a small group (five) of Sphoeroides testudineus fish, normally on the standard Seaquarium diet of shrimp and blue runner, was fasted for 48 hr, and then 50 gm each of Bryopsis hypnoides, Gracilaria compressa, and Ulva lactuca were placed before them. None appeared to be eaten during an observation period of an hour. On feeding the standard diet (shrimp and blue runner), consumption started at once.

III. RESULTS

The results achieved by these feedings, examinations, and injections are presented in Tables I - IX.

IV. DISCUSSION

The toxicity of the algae native to the waters in which toxic Sphoeroides nephelus (Lalone et al., 1963) were common was found to be very low (Tables I and II). There was only one death in the whole series (63 feedings) of hamsters (Table I), and this was due to aspiration of the suspension during the forced feeding procedure, as confirmed by lung pathology at autopsy. Neither Bryopsis hypnoides nor Gracilaria compressa were toxic to hamsters (Table II), but their intraperitoneal injections into mice produced some late deaths (Table II). These deaths did not take place during the first hour, typical of Sphoeroides toxicity (Larson et al., 1967), but were delayed, occurring during the second or third day. Gorham (1964) has ascribed delayed deaths from algal extracts to bacterial contamination. Our work seemed to confirm this, since bacteria-

* Help on classification and identification was given by Dr. Donald R. Moore, Rosenstiel School of Marine and Atmospheric Sciences, University of Miami.

TABLE I

Hamsters: Oral Administration

Algae	Number of feedings	Number showing no apparent effect
Bryopsis hypnoides	24	24
Gracilaria compressa	16	16
Ulva lactuca	23	22 [a]

[a] One nonsurvivor: pneumonia.

TABLE II

Mice: Intraperitoneally Injected

Algae	Filtered extract		Bacteria-free extract	
	Number of animals	Number surviving [a]	Number of animals	Number surviving
Bryopsis hypnoides	20	18	20	20
Gracilaria compressa	22	18	20	20
Ulva lactuca	14	14	-	-

[a] Deaths during the second or third day.

free algal extracts (Table II) produced no delayed deaths. The Ulva lactuca was found to be nontoxic to either mice or hamsters. Therefore, bacteria-free extract of Ulva lactuca (Table II) was not prepared because the original extract was nontoxic.

No algae were found in the gastrointestinal tracts of any of the fish examined, either the S. nephelus or S. testudineus. Sphoeroides nephelus have been observed (E.L.) browsing on algae-covered concrete bridge abutments or the concrete slabs protecting the fill on the causeway between Cocoa and Cocoa Beach, Florida, but perhaps other organisms living among the algae, or on which algae had grown, were sought for food instead of the algae. Bryopsis (sp. not stated) have been found by Habekost et al.,(1955) to be weakly toxic by the mouse injection test and Halstead (1967) states: "It is believed that toxic algae also may be a factor in toxin production in tetradontiform fishes." Freshwater blue-green algae contain a toxin similar to saxitoxin (paralytic shellfish toxin))but different from tetrodotoxin (Jackim and Gentile, 1968).

The results with the algal feeding to chicks show that the weight increases (Table III) are similar. Some difficulty was evident at the start to induce the chicks to eat this diet of half-algae and half-chick food in spite of a sprinkling of the pure Purina Startena (on which they had been raised) on top of the mixture. Though we could not be certain that all three groups received the same isocaloric intake the average weights recorded are in accord with published data (Altman and Dittmer, 1964). From quantitative analysis of six species of marine algae (one Gracilaria and one Ulva) from Puerto Rico and the Virgin Islands, Burkholder et al. (1971) believe that the essential amino acids found in algae could improve the food chains in the production of fish. Bardach (1968), in reviewing aquaculture procedures, reports the commercial raising of milkfish on a diet containing blue-green algae.

TABLE III

Chicks Feeding (Weights in Grams)

Algae	Number of animals	Start	Middle	End
Bryopsis hypnoides	10	145	182	233
Gracilaria compressa	10	178	217	257
Ulva lactuca	11	219	230	277
Lactuca sativa [a]	10	227	277	321

[a] Control.

The temperature variations (Table IV) seem negligible and eliminate the presence of either pyrogenic or depressant substances in the algae. The temperatures recorded correspond with the observations of others (Wilson and Plaister, 1951). With

TABLE IV

Chick Temperatures (Average Degrees, Centigrade)

Algae	Number of animals	Start	Middle	End
Bryopsis hypnoides	10	42.1	41.9	41.9
Gracilaria compressa	10	42.1	42.1	42.1
Ulva lactuca	11	41.7	41.9	41.9
Lactuca sativa [a]	10	41.9	41.9	42.0

[a] Control.

exception of the control group, there was a decrease of the respiratory rate during the experimental period. This can be attributed to conditioning and also their greater weight. From their physical appearance during the entire experimental period, all these chicks appeared normal, their combs and wattles developed, feathering progressed, and some became pugnacious. In all respects they seemed the same as the control group fed Purina Startena and ordinary leaf lettuce (Lactuca sativa). Our results with the feeding of chicks with algae as well as the negative mouse injection tests and hamster feedings certainly indicate, if not prove, that the algae, Bryopsis hypnoides, Gracilaria compressa, or Ulva lactuca, are not the source of toxicity of Sphoeroides nephelus or Sphoeroides testudineus.

These two species of toxic puffers were taken from distinct areas about 200 miles apart. Most of the S. nephelus were caught in the shallow waters along the causeways of the Banana River, near Cocoa, and in the Indian River, near Titusville, Florida. The S. testudineus specimens were caught in shallow and medium depth waters near Baker's Haulover or at Turkey Point, Miami, Florida. Control fishes (grunts, snappers, jacks, and catfishes) were obtained from these three areas. Recently, Burklew and Morton (1971) have found that S. nephelus from the eastern Gulf of Mexico were nontoxic by the mouse injection test.

Examinations of the gastrointestinal tracts of both S. nephelus and S. testudineus under a low power microscope showed that both contained two species of Brachidontes and two species

TABLE V

Chicks Respiratory Rates (Average per Minute)

Algae	Number of animals	Start	Middle	End
Bryopsis hypnoides	10	72	60	64
Gracilaria compressa	10	64	52	58
Ulva lactuca	11	62	59	57
Lactuca sativa [a]	10	50	53	52

[a] Control.

TABLE VI

Gastro Intestinal Examination:
Number of Fish, Showing Mollusk Content,
Either Whole Shell or Fragments

Species	*Sphoeroides nephelus* 38	*Sphoeroides testudineus* 87
Brachidontes exustus	28	20
Brachidontes citrinus	11	9
Chione cancellata	10	5
Chione grus	7	7

of *Chione* (Table VI). Other mollusks also were present (Table VII) though in smaller numbers. Also included in the GI tract contents were small whole crabs and crab fragments, small shrimps, some bits of sea grasses, and sand. The control fishes had no mollusks in their gastrointestinal tracts. Our findings are similar to those of the comprehensive study of food habits of the reef fishes from the West Indies (Randall, 1967).

The coincidental findings of shell forms held in common to both species of puffers in this study may contribute to an explanation of their acute toxicity. It cannot be overlooked

TABLE VII

Gastro Intestinal Examination:
Mollusks found in 20 Sphoeroides nephelus Fish
from Banana River, Cocoa, Florida

Species	Average number per fish
Brachidontes citrinus	7
Brachidontes exustus	4
Chione cancellata	2
Chione pygmaea	2
Nassarius vibex	1 - 2
Nassarius ambiguus	1
Anomalocardia cuneimeris	1

that the two groups were obtained from regions that were more than 200 miles apart, yet the skin of both species was quite toxic and both consumed four of the same types of bivalves, i.e., characteristic of their eating habits. Brachidontes exustus and Chione cancellata administration to the chick in large doses, 5% of the body weight, produced no discernible effects (Table VIII). The chick is a susceptible animal to puffer fish toxins (Larson et al., 1967).

If mollusk content was related to the toxicity of these two species, S. nephelus and S. testudineus, they should have the same or nearly the same toxicity. Table IX shows the relative toxicities of the skin and flesh from these two species, the skin of both being highly toxic by the mouse injection test, only one mouse in each large series surviving. When the

TABLE VIII

Mollusk Slurry Fed to Chicks

Species	Number of animals	No apparent effect
Brachidontes exustus	22	22
Chione cancellata	21	21

TABLE IX

Comparative Toxicity of Sphoeroides nephelus and Sphoeroides testudineus:

Ratio of number of animals tested to deaths, N/D. The IP injections were 1 ml/20 gm body weight

	Mice IP		Feeding 5% of body weight	
Tissue	S. nephelus	S. testudineus	S. nephelus	S. testudineus
Skin	46/45	51/50		
Flesh	40/40	44/7	15/13	54/0

toxicity of the flesh was determined by the mouse injection test, the situation differs considerably. The S. nephelus group had no survivors (40/40), whereas with S. testudineus only 7 of the 44 mice died. The flesh feeding tests with chicks also are very significant since the LD_{50} for oral administration is about 1.25% of body weight for S. nephelus (Larson et al., 1967), whereas the oral consumption of 5% of S. testudineus produced no deaths, but in a few of the chicks the legs were incoordinated for about an hour.

The Tetraodon or puffer poison may not be dependent on the diet of the fish. The algae do not seem to be the source since they showed no toxicity by three procedures (Tables I – V) and also were not found in the gastrointestinal tracts. The presence of the mollusks in the gastrointestinal tracts of both toxic species, as well as the absence of toxicity in other fish in the area (catfish, etc.) whose gastrointestinal tracts do not contain mollusks, might indicate that mollusks are the source or precursor of the toxic principle or principles present in S. nephelus. However, we believe there is a better explanation: The Tetraodontidae have powerful jaws and heavy teeth suitable for crushing mollusks and similar food, whereas the other fish lack these adaptations. Halstead (1967) in his excellent review quotes several ideas for the origin but states: "puffer poison is not a decomposition product but a normal component of the living fish." Taft (1945) and also McFarren (1967) have stated that the toxin is endogenous, i.e., produced by metabolic activity of the organism itself. With these latter statements, the authors agree and could conceive a biosynthesis of tetrodotoxin, a quinalzine compound (Mosher et al., 1964), since Florkin and Stotz (1968) and also Bernfeld (1967) have stated that quinalzine compounds can be formed from phenylalanine (an essential amino acid).

V. SUMMARY

The algal theory for the production of toxicity in the puffer fishes, S. nephelus and S. testudineus, seems untenable because of negative evidence from three different types of experiments. The theory that mollusks could be the source of toxicity seems also to be weak for the following reasons: (1) absence of mollusks in some specimens, (2) nontoxicity with feeding experiments, and (3) the difference in toxicity between the two species of Sphoeroides with similar mollusk intake.

From our results, we conclude that the toxin of Sphoeroides is endogenus, being a metabolic product.

REFERENCES

Abbott, R.T., "American Seashells." Van Nostrand, New York, 1953.

Altman, P.L., and Dittmer, D.S., "Biology Data Book," Federation of American Societies for Experimental Biology, Washington, D.C., 1964.

Bardach, J.E., Science 161, 1098 (1968).

Bernfeld, L., "Biosynthesis of Natural Compounds." Pergamon Press, Oxford, 1967.

Burkholder, P.R., Burkholder, L.M., and Almodovar, L.R., Botanica Marina 14, 132 (1971).

Burklew, M.A., and Morton, R.A., Toxicon 9, 205 (1971).

Florkin, M., and Stotz, E.H., "Comprehensive Biochemistry," Vol. 20. Elsevier, Amsterdam, 1968.

Gorham, P.R., "Toxic Algae." (C.F. Jackson, ed.). Plenum Press New York, 1964.

Habekost, R.C., Fraser, I.M., and Halstead, B.W., J. Wash. Acad. Sci. 45, 101 (1955).

Halstead, B.W., "Dangerous Marine Animals," Cornell Press, Cambridge, Md., 1959.

Halstead, B.W., "Poisonous and Venomous Marine Animals," Vol. II, pp. 712- 713, U.S. Government Printing Office, Washington, D.C., 1967.

Jackim, E., and Gentile, J., Science 162, 915 (1968).

Lalone, R.C., de Villez, E.J., and Larson, E., Toxicon 1, 159 (1963).

Larson, E., Lalone, R.C., de Villez, E.J., and Siman, R. Jr., in "Animal Toxins" (F.E. Russell and P.R. Saunders, eds,). Pergamon Press, Oxford, 1967.

McFarren, E.F., in "Animal Toxins" (F.E. Russell and P.R. Saunders, eds.), Pergamon Press, Oxford, 1967.

Mosher, H.S., Fuhrman, F.A., Buchwald, H.D., and Fischer, H.G., Science 144, 1100 (1964).

Randall, J.E., Proc. Int. Conf. on Trop. Oceanography V. 665 (1967).

Taft, C.H., Texas Reports on Biology and Medicine 3, 339 (1945).

Tressler, D.K., and Lemon, J. McW., "Marine Products of Commerce," Reinhold Publishing Co., New York, 1951.

Williams, R.H., Educational Series, Florida State Board of Conservation No. 7, 1950.

Wilson, W.O., and Plaister, T.H., Amer. J. Physiol. 166, 572 (1951).

Chapter 8

SOME PHYSIOLOGICAL PROPERTIES OF DINOFLAGELLATE TOXINS

Dean F. Martin

Department of Chemistry, University of South Florida
Tampa, Florida

George M. Padilla

Department of Physiology and Pharmacology
Duke University Medical Center
Durham, North Carolina

I. INTRODUCTION

A primary interest in flagellate toxins arises from the involvement of the organisms in "red tides" and attendent phenomena. The term "red tide" is a general one applied to areas of discolored water that arise from the sudden and sporadic development of extremely large numbers of flagellates (see also Collier, 1971). Attendent phenomena include catastrophic fish kills followed by serious economic and public health problems. Although the flagellates may be present at all times in small numbers as a component of the normal plankton population, the precise combination of water conditions that stimulates production of red tide is still not known.

This is but one of many aspects of red tides about which comparatively little is known. Many believe that the conditions that prolong the massive bloom are generally poorly defined, though Ryther (1955) made the following suggestion:

> ''there is no need to postulate obscure factors to account for prodigious growth of dinoflagellates...it is necessary only to have conditions favoring the growth and dominance of a moderately large population of a given species, and the proper hydrographic and meteorological conditions to permit the accumulation of organisms at the surface and to effect their further concentration in localized areas.''

Moreover, we should like to know the materials that are produced by phytoplankton during blooms and the interrelationships of other intra- and ultra-bloom species. Finally, the function, if any, of toxins within phytoplankton has been a matter of considerable curiosity.

Economic and public health problems arise from two effects, primary and secondary. The former is due to a toxin that is present in or is released by the flagellate. A secondary effect, oxygen depletion, may also be responsible for castrophic deaths of invertebrates. In the latter instance, any algal bloom can cause oxygen depletion through respiratory activity in itself or through intense bacterial activity associated with decomposition of the bloom organism.

Two kinds of primary effects, i.e., those associated with a toxigenic organism, are observed: direct (respiratory failure) and indirect (e.g., paralytic shellfish poisoning). The direct primary effect arises from susceptible animals being subjected to lethal concentration of a toxin that initially paralyzes them and then leads to death through respiratory failure. The indirect effect arises through the food chain. Shellfish (typically mussels and clams) on exposure to the toxigenic organism (such as Gonyaulax catenella and G. tamarensis) mechanically entrap the toxin and are rendered toxic to higher organisms, the most prominent of which is man. Over 200 human fatalities have been attributed to paralytic shellfish poisoning (Halstead, 1965).

The toxin effects of phytoplankton vary in their kind of action. For example, members of the genus Gonyaluax generally seem to be implicated only in paralytic shellfish poisoning, and fish kills if they arise seem to come from the secondary effects of oxygen depletion. In contrast, Gymnodinium breve has been implicated in massive fish kills (cf. Rounsefell and Nelson, 1966), shellfish toxicity (McFarren et al., 1965; Cummins et al., 1971), as well as massive deaths due to the secondary effects (Steidinger et al., 1973).

The primary toxic effects of phytoplankton also vary in intensity. For example, symptoms and signs of shellfish poisoning are relatively mild and in man include loss of equilibrium, diarrhea, gastrointestonal distress, and tingling sensations (around mouth, face, and extremities) leading to mild paralysis (McFarren et al., 1965). Eye irritation and respiratory problems have also been reported by persons who stood on the shore near blooms, as well as some distance away (cf. Ingle, 1954; Woodcock, 1948; Hutton, 1956). No fatalities due to paralytic shellfish poisoning because of G. breve have been reported. In contrast, the effects of paralytic shellfish poisoning due to saxotoxin (from G. catanella) are much more severe. Human fatalities have been observed, and the IV LD_{50} value is 3.4 μg/kg of body weight in the mouse (Wiberg and Stephenson, 1960) vs. about 500 μg/kg for strychnine. Mussel poisoning symptoms in man were summarized by Schantz (1961), who noted that the symptoms usually developed within 0.5–3 hr after consuming mussels and that the symptoms included numbness (lips and fingertips) followed by paralysis and finally death by respiratory failure. Death occurred within 3–20 hr, depending on the dose, and survival beyond 24 hr indicated a favorable prognosis.

Primary effects have been suspected for over 20 dinoflagellates in fresh and marine environments (Halstead, 1965). In several instances, red tides and the source of toxins has been conclusively associated with specific organisms. The marine organisms include Gonyaluax catenella (Central-North Pacific coasts, United States and Canada, Burke et al., 1960; Schantz, 1970), Gonyaulax monilata (Gulf of Mexico, Connell and Cross, 1950), Gonyaluax acatenella (Strait of Georgia, British Columbia, Prakash and Taylor, 1966), Gonyaulax tamarensis (Atlantic Coast of Canada, especially the Bay of Fundy, Prakash, 1967), Gymnodinium breve (Gulf of Mexico, Ray and Wilson, 1957; Martin et al., 1972, 1973), Gymnodinium venificum (British waters, Abbott and Balantine, 1957), and Prymnesium parvum (Europe and Middle East, brackish-to-saline waters, Shilo, 1967).

Three of the organisms mentioned have received intense study. The physiological properties of saxitoxin are reviewed by Kao (Chapter 5). The chemistry and chemical environment of G. breve has been described elsewhere (Collier, 1971, Rounsefell and Nelson, 1966; Wilson, 1965; Martin and Padilla, 1971a,b; Steidinger et al., 1973), as have the characteristics of the toxins (Cummins et al., 1971; Martin and Chatterjee, 1970; Trief et al., 1970). The characteristics of prymnesin, the general toxin obtained from Prymnesium parvum, will be reviewed in subsequent sections, together with some techniques for characterizing this toxin and others in terms of hemolytic activities and membrane-binding tendencies.

II. PRYMNESIUM PARVUM TOXINS

Over 30 years ago mass fish mortalities were first linked to blooms of the phytoflagellate Prymnesium parvum. Many investigations have been performed to identify and isolate the toxic principles responsible for such fish kills (Otterstrøm and Steemann-Nielsen, 1939, as cited in Shilo and Rosenberger 1960). At least two types of toxins are believed to be produced by Prymnesium parvum, as shown by a differential ichthyotoxicity and hemolytic activity. In addition, a distinct neurotoxic action has been ascribed to P. parvum toxins (see Shilo, 1967 for review). It is difficult, however, to ascribe any of these toxicities apart from the impairment of membrane formation caused by this poison. A notable exception is the interference of P. parvum toxins with the action of neurotransmitters such as acetyl choline (Bergmann et al., 1964; Parnas and Abbott, 1965).

Specifically, P. parvum toxins inhibit the acetyl choline-induced contraction in the intestine (Bergmann et al., 1964). Since more than one toxin may be present, none of which has been purified or characterized in a strict chemical sense, the results are open to many interpretations, a situation which is common with most biotoxins of marine origin.

Investigators are thus forced to follow two main avenues of experimentation. While the chemical and physical nature of the toxin is being defined with improved methods of isolation and purification, new bioassays must likewise be devised to separate a specific biological activity unequivocally associated with the pure compound. Even if this ultimate goal is not achieved, much can be learned about a toxic principle if it interacts with known cellular components and brings about specific changes in cell function as do compounds with clearly defined chemical and biological characteristics. For example, Ulitzur and Shilo (1970a) drew a strong similarity between the action of prymnesin and that of surface-active agents such as sodium dodecyl sulfate and cetyltrimethylammonium bromide. Rauckman and Padilla (1970) and Martin and Padilla (1971a,b) examined the hemolytic activity of prymnesin, its possible role in membrane stabilization, and its interaction with exogenous phospholipids and cholesterol.

A. Ichthyotoxins

Much of the early work arose from ecological and economic considerations (Reich and Aschner, 1947, as cited in Shilo and Shilo, 1953). The initial impetus was directed toward evaluating the ecological, nutritional, and physical factors in the production of toxin, both in the field and laboratory (McLaughlin, 1958; Parnas and Spiegelstein, 1963; Shilo and Rosenberger, 1960). As early as 1947, research Workers (in Israel in particular) had developed methods for the qualitative bioassays of the toxin, the major one being based on the lethality of the toxin toward fish (see Shilo, 1967). A quantitative ichthyotoxin bioassay was ultimately worked out in detail by Bergmann et al., (1964), An improvement over the previous methods involved the injection of the toxin into specimens of the fish Gambusia affinis. A unit of toxin was defined as the quantity which produces 50% mortality (LD_{50}) in Gambusia minnows of 300 mg average body weight following intraperitoneal injection (Bergmann et al., 1964). Evidence was presented to show the injection technique gave very reproducible results and was much more sensitive than the "immersion technique."

With the development of axenic cultures of P. parvum, investigations on the chemical nature of the toxins released by cells grown in culture, the factors which influence their formation, and the pharmacological activity of toxins soon followed (see Parnas, 1963; Shilo, 1967, for excellent reviews of this work). For example, Shilo and Rosenberger (1960) studied the biological activity of toxins formed by P. parvum and also evaluated factors important in the biosynthesis of the two major toxic principles. Their studies included an investigation of the lysis of human cells by P. parvum toxins. Up to this point much attention was directed toward the study of the extracellular toxins recovered from the culture medium by extraction with methanol (Yariv, 1958, as cited in Shilo and Rosenberger, 1960). The toxin(s) (named prymnesin) was then concentrated by adsorption on magnesium hydroxide precipitates (Yariv and Hestrin, 1961).

Shilo and Rosenberger (1960) found that greater quantities of intracellular toxins could be extracted from the cells with methanol. Although a quantitative assay for ichthyotoxicity proved difficult, they found that much greater amounts of hemolysin were obtained than ichthyotoxin. The authors were careful to note that the two toxins resembled each other in their solubility in methanol but not acetone, were nondialyzable, and were not retained by cation- or anion-exchange resins. While both toxins were readily inactivated by moderately elevated temperatures (70°C), only the hemolysin was readily destroyed by brief exposure to 0.5 N NaOH for short periods (60 sec). This was one of the early indications that several toxins are produced by P. parvum, but as recently stated by Shilo (1967):

"The difficulties in separation and the close chemical properties of the different active fractions seem to indicate that all the P. parvum toxins belong to a closely related family of chemical substances. Since homogeneity of even our most potent *Prymnesium* preparations has not yet been established beyond doubt, it seems premature to draw more detailed inferences concerning the chemical nature of the toxin....''

In spite of the experimental limitations underlined by the above statement, Ulitzur and Shilo (1964, 1966) conducted a detailed series of investigations on the ichthyotoxic fraction. These studies are of particular interest in giving us some insight on the cellular basis for this toxicity. They studied the uptake of Trypan blue and radioactive-labeled human serum albumin as a means of determining the damaging effect of the ichthyotoxin on gill membranes.

Ulitzur and Shilo (1964, 1966) demonstrated that damage to the gills induced by brief exposure to ichthyotoxin is reversible. They concluded that intoxication of gill-breathing animals takes place in two stages: First, a reversible damage leads to the loss of selective membrane permeability, under the specific conditions required for ichthyotoxic activity (i.e., presence of cationic synergists and a suitable pH, about 9); and second, fish mortality results from the sensitization of the fish to various "toxicants" present in the surrounding medium, which, though not toxic when added directly upon the gills (in the absence of prymnesin), can now penetrate the membrane, enter the systemic circulation, and induce paralysis and death.

In summary, the primary effect of the toxin was to increase the gill membrane permeability. Although it is not clear from available data how selective this change was, the ultimate passage of labeled serum albumin would indicate that the damage was considerable. It is clear, however, that the damage occurred gradually, reflecting either the complexity of the gill membranes or a response determined by the dosage. It is interesting to note that the kinetics of toxicity in the fish-kill test and isolated guinea pig preparation were of the same order of magnitude, namely, approximately 10 min are required for the full effect of the toxin to be manifested.

In view of the above considerations and the well-documented observation that only gill-bearing animals are poisoned by *P. parvum* extracts (Shilo, 1967), it is expected that the toxic effect will be explainable in terms of membrane disfunctions of one type or another. This becomes more apparent when we consider the other two toxic principles: cytotoxin and hemolysin. It would seem to the present reviewers that these terms are interchangeable, merely reflecting the cell types on which the toxic principles are applied. However, until this proves to be the case, it is best to consider cytotoxins apart from the hemolysins.

B. Cytotoxins

Prymnesin is toxic to tissue culture cells. Shilo and Rosenberger (1960) were the first to find that the sensitive cell lines included Ehrlich ascites cells, HeLa cells, human amnion cells, and the Chang strain of liver cells. A common feature of the cytotoxic action was the induction of pseudopodialike extrusions. Ultimately complete breakdown of the cell membrane occurred as indicated by the uptake of the dye Trypan blue. The hemolytic and ichthyotoxic activity in various preparations of prymnesin varied (as much as 700 to 1), an early indication that at least two different toxic factors were produced. Another clue: the observation that the hemolytic titer could be reduced by as much as 90% on exposure of the hemolysin to alkaline methanol. No attempt was made in these early investigations to compare the cytotoxic activity between tumor and normal cells.

Subsequently, Dafni and Shilo (1966) studied events leading to cytolysis of Ehrlich ascites tumor cells by P. parvum cell extracts. Four criteria were used defining cytotoxic action: (1) assimilation of Trypan blue by the cells; (2) release of intracellular macromolecules (RNA and protein); (3) change in cell volume of the injured cells; and (4) microscopic observations of morphological changes. The morphological changes induced by the cytotoxins were similar to those described by Shilo and Rosenberg (1960).

Kinetic analysis of cytolysis (Dafni and Shilo, 1966) revealed that prior to cytolysis a marked increase in the median volume of the cells occurred immediately on addition of the cytotoxin. Yet the time course and extent of swelling were dependent on pH and temperature, as was the loss of material from the swollen cells. For example, at pH 6.4 and 27°C the cells showed marked swelling, but leakage of macromolecules or uptake of the dye occurred. At pH 7.4 and 37°C, however, the cells were readily stained and lost considerable cytoplasmic material. It was only when the cells had reached a maximum level of swelling that they began to lose material, at which time the cell volume decreased drastically.

Dafni and Shilo (1966) suggest that the sequence of lysis elicited by prymnesin is similar to that seen in immune cytolysis. In a subsequent report, Dafni (1969) made use of the fact that at a lower pH (6.4) only swelling is observed and found that within the first 3-10 min, 80-90% of the potassium leaked out of the cells with an apparent exchange with external sodium. No change in the cell volume was seen at this point. Further incubation resulted in a continual entry of external sodium but at a slower rate, but since the potassium content was constant, a gradual swelling ensued. The primary effect of P. parvum cytotoxin was due to a disruption of the selective membrane permeability which appears to be pH independent, according to Dafni (1969).

The effect of P. parvum toxins on several bacteria and mycoplasma organisms was studied by Ulitzur and Shilo (1970a). Intact cells were unaffected by the toxin at a concentration of 3.3 μg/ml (using the intracellular fraction "Toxin B" purified by the method of Ulitzur and Shilo, 1970b). Intact bacteria may be resistant to P. parvum toxins primarily because the cell wall prevented entry of the toxin and thus nullified its attack upon the bacteria membranes, according to the authors, because removal of the cell wall (by EDTA – lysozyme treatment) increased the sensitivity to the toxin (Shilo, 1971).

A unique chemical structure of P. parvum toxin preparation may be responsible for the strong affinity for biological membranes. On the basis of published chemical analyses (Paster, 1968) and isolation by various purification schemes [column chromatography, thin-layer chromatography, gel electrophoresis, etc. (Ulitzur and Shilo, 1970b)], the toxin appears to contain lipid and polar moieties. In this respect they are comparable to synthetic detergents and lysophospholipids, though P. parvum toxin is at least 3000 times more active than digitonin or isolechitin in hemolytic potency (Ulitzur and Shilo, 1970b). Thus both the hemolysin and cytolysin may belong to the same family of toxin principles, which affect cells by attacking the membrane.

C. Hemolysins

A wide variety of biological activities associated with membrane function and the methods have been developed to measure them (e.g., determination of ion transport by isotopic techniques, measurement of ionic conductances, antibody-antigen interaction at the cell surface, contact inhibition, etc.) Curiously, the technique of hemolysis has been relegated to a secondary position as an analytical method in recent years.

Probably because of its inherent simplicity it has been mainly used as a means of assessing the strength and purity of toxins derived from P. parvum.

Reich et al. (1965) described the hemolytic action induced by P. parvum toxins and reported that hemolysis began with a delay of about 15 min; the maximal lytic effect was reached after about 40 min. The rate of hemolysis (reflecting decreases in the "latent" period))was dependent on the concentration of toxin adsorbed on the red cells and therefore proportional to the external concentration. Even a short exposure to the toxin, however, was sufficient for binding. For example, sequential batches of erythrocytes were exposed to the toxin and the hemolytic activity remaining in the supernatant fluid was measured by exposing "fresh" cells to it. Up to 80% of the toxin could be removed by four separate exposures of red blood cells during the latent period without their undergoing hemolysis.

Bergmann and Kidron (1966) examined further the relationship between toxin binding and hemolysis using rabbit erythrocytes and Ehrlich ascites cells. Bacterial β-hemolysin, added at a concentration that did not induce hemolysis in rabbit erythrocytes, was readily absorbed by the cells and induced a "prolytic" change: The lag phase of subsequent osmotic lysis was reduced progressively from approximately 150 to 40 sec when the cells were exposed up to 20 min with 0.05 IU/ml of β-hemolysin. Beyond 20 min no further reduction in the prolytic phase occurred. Pretreatment of rabbit erythrocytes with β-hemolysin had a synergistic effect on prymnesin-induced hemolysis by increasing the level from approximately 6% (control) to 53%. Very small quantities of lytic agent could be detected, and once the binding had taken place, washing did not remove it from the cell.

By combination of two hemolytic processes, latent disruptive effects of lytic agents can be uncovered and evaluated quantitatively. In these studies, the synergistic effect of β-hemolysin and prymnesin was mainly analyzed in terms of the reduction of the latent (or prolytic) phase of hemolysis and subsequent loss of hemoglobin (expressed as the titer of hemolysis). The initial effect of β-hemolysin was presumed to be due to an enzymatic interaction with the sphingomyelin (or related components) of the membrane. Since erythrocytes from different animals have different types and proportions of phospholipids, differential effects can be obtained with a given lytic agent. Yet, differential sensitivity can also result from a change in arrangement of the phospholipids within the membrane, or their increasing exposure to the lytic agent as the cell is osmotically deformed during prolytic phase and early stages of hemolysis. Thus, we saw an obvious need to examine more closely the effect of P. parvum toxins on the kinetics of hemolysis and relate this phenomenon to specific membrane functions in erythrocytes. The methods and experimental protocols employed in these studies will be described in some detail below.

III. METHODS

A. Materials

1. Isotonic Buffer

Isotonic buffer, 10 mM phosphate-buffered saline solution, pH 5.5, has been used commonly to study the hemolytic activity of various lysins. Another common buffer, Ringer's blood buffer, has also been used.

2. Blood

Heparinized rabbit blood, stored at $4^{o}C$, was suspended in isotonic buffer. Erythrocytes were collected by centrifugation in a clinical centrifuge (2 min at 1800 rev/min) and were washed twice with blood buffer. For determination of the hemolytic activity for kinetic studies, a standard erythrocyte suspension was prepared by dilution to an absorbance of 1.00 unit at 540 nm. Under these conditions, the electronic cell count of the standard suspension was about 10^7 cells/ml. When completely lysed, the standard cell suspension had an absorbance of about 0.18–0.20 units at 540 nm.

3. *Prymnesium parvum* Toxin

The toxin was isolated from *P. parvum* cultures grown in artificial seawater medium (10 parts per thousand), maintained at $25^{o}C$ under constant illumination (Padilla, 1970). The seawater (or an artificial seawater mixture such as Instant Ocean, Wycliffe, Ohio) is enriched with liver fusion (Oxoid, Inc., 300 mg/liter, glycerol (0.25M), and vitamins B_1 and B_{12} (0.02 and 0.01 μg/liter, respectively). After 5–7 days of growth at a density of 10^7 cells/ml, cells are collected by centrifugation (5,000 g, 10 min), ground in a tissue grinder, and extracted of pigment with acetone. The acetone and insoluble residue was extracted several times with methanol to obtain the toxin. The combined methanolic extracts were purified by gel-filtration chromatography using Sephadex LH-20 in methanol (Martin and Padilla, 1971a).

4. Labeled Prymnesin

Toxins were obtained from cultures enriched with suitable radioactive precursor materials. For example, artificial or natural seawater (500 ml) was first inoculated with 10-20 ml of glycerol-free P. parvum stock culture. The inoculated culture was then treated with presterilized (e.g., by filtration through 0.22 μm porosity Millipore filters) 30 ml of 3.44 M glycerol containing 500 μCi of [2-^{3}H] glycerol (200 mCi/mM, New England Nuclear Corp). Cells were harvested, and the toxin was isolated and purified as described above (cf. Martin and Padilla, 1971a).

B. Techniques of Hemolysis

1. Hemolytic Units

Hemolytic activity was determined in two ways: as an extent of hemolysis or by kinetic determination. The first method is commonly used to determine the hemolytic units present. A hemolytic unit is defined as that amount of toxin in 0.1 ml of methanol that causes 50% hemolysis of 2.9 ml of standard erythrocyte suspension after a defined time at a defined temperature. The defined conditions do vary from worker to worker, e.g., Padilla and co-workers (Padilla, 1970; Martin and Padilla, 1971a; Rauckman and Padilla, 1970) determined hemolytic activity at room temperature, 10 min, and pH 5.5, the pH at which hemolytic activity was found to be maximal for rat erythrocytes (Padilla, 1970). [See Reich et al. (1965) and Paster and Abbott (1969) for other defined conditions.]

2. Determination of Kinetics

The second way in which the hemolytic potency of toxins were assayed was by establishing the rates of hemolysis colorimetrically at 540 nm using a double beam spectrophotometer (e.g., Beckman DB-G), equipped with a thermostatted cell compartment and a strip-chart recorder. The reference cell was filled with blood buffer, and the sample cell was filled with 2.9 ml of standard erythrocyte suspension, which was allowed to reach reaction temperature before being mixed thoroughly with 0.1 ml of methanolic toxin solution. Kinetic data (absorbance as a function of time) were obtained from the strip-chart recorder, directly, if it is equipped with a linear-log attachment, or indirectly as percent transmittance (Fig. 1), being

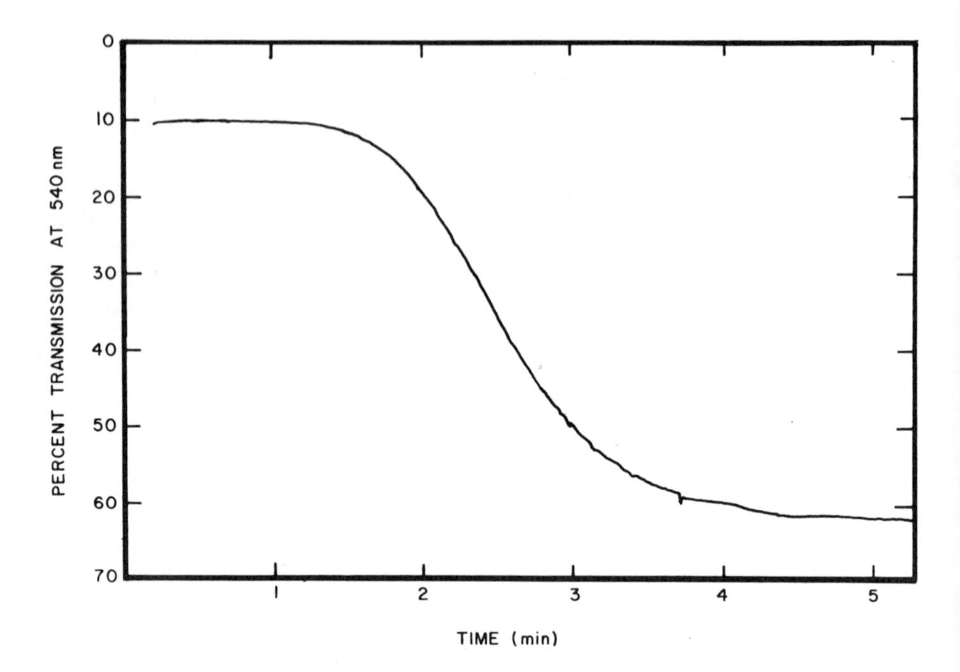

FIG. 1. Typical rate plot of hemolysis of rabbit erythrocytes using 2.5 hemolytic units of prymnesin at 25°C and pH 5.5. (From Martin and Padilla, 1971a, with permission of Elsevier Publishing Co.)

converted by means of calibration of absorbance values. Kinetic parameters of a useful sort can be obtained directly from a plot mechanically, calculated by conventional means, or obtained as computer-derived constants, from a spectrophotometer equipped with a digitizing unit. The effect of inhibitors was measured essentially the same way except that a methanolic solution of toxin (0.05 ml) and inhibitor (0.05 ml, 6 mM) was added with mixing to 2.9 ml of RBC suspension.

C. Kinetic Analysis

Three groups of parameters were derived from the conventional colorimetric kinetic determinations. These include qualitative parameters, first-order kinetic parameters, and computer-derived parameters. These may be considered in order in the following subsections.

1. Qualitative Parameters

From the plot obtained [absorbance as a function of time (Fig.2)], we see that the changes associated with hemolysis can be divided into three phases: a prolytic period, the initial phase preceding the onset of the lysis during which there is very little change or perhaps a slight increase in absorbance associated with cell swelling; the lytic phase, characterized

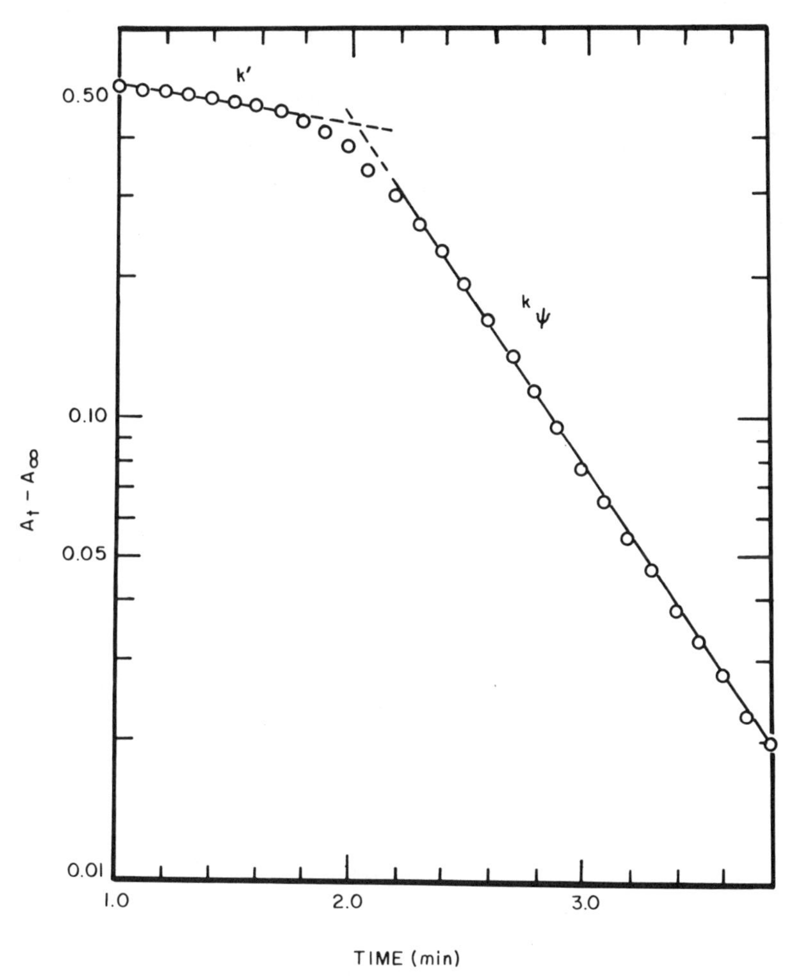

FIG. 2. First-order rate plot of prymnesin-induced hemolysis using data obtained from Fig.1. (From Martin and Padilla, 1971a, with permission of Elsevier Publishing Co.)

by rapid change followed by the final phase; an absorbance independent phase. The first and last phases can be characterized in general terms as by the length of time of prolytic period, t_p, which has been used previously (Martin and Padilla, 1971b), and the final phase can be characterized by the value of A_∞, the absorbance of the final solution. Qualitatively, t_p varied inversely with the concentration of hemolysin and directly with the concentration of inhibitor present (Martin et al., 1971a). The value of A_∞ typically was about 0.2 for the conditions defined earlier, but the value is also a function of the concentration of inhibitor present and will increase, indicating incomplete reactions with increasing concentrations of inhibitor. In addition, the reciprocal of the length of the prolytic phase, t_p^{-1}, was related directly to concentration of certain inhibitors. These values, reflecting as they do the general phases of the course of the onset of hemolysis, are useful and have the advantage that they may be obtained from a direct chart reading. A third parameter, initial slope of the kinetic phase, $(\Delta A/\Delta t)$ might be used but it appears that obtaining kinetic data as first-order rate constants, for example, may be a more useful practice.

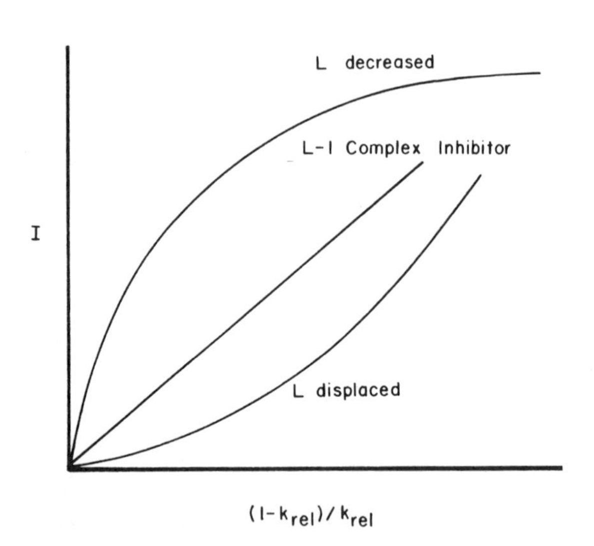

FIG. 3. Diagnostic plot, inhibitor concentration as a function of the inhibition parameter, showing relationships expected for three pathways of inhibition of prymnesin-induced hemolysis.

2. First-Order Kinetic Constants

The prolytic and lytic periods were characterized in a more satisfactory manner by calculating specific rate constants. Consecutive pseudo first-order rate constants, k' and k_ψ were obtained from conventional first-order rate plots (Fig.2). It was somewhat difficult to obtain accurate values for the first-rate constant that appear to be associated with toxin binding and cell swelling during the prolytic period just prior to the onset of the lytic phase. The accuracy of k' was limited by the number of points available for any satisfactory estimate, at least at room temperature, and by the need for an accurate estimate of A'_∞, the final absorbance of the first process. The specific rate constants associated with the second process, k_ψ, could be obtained with good precision and usually good first-order rate plots were obtained for better than 95% of the reaction. The constants were calculated from the linear portion of the rate plot $-\log_e(A_t - A_\infty)$ as a function of time using the conventional relationship $k_\psi = m$. Here m is the slope, and A_t and A_∞ refer to the absorbance values at times t and at the completion of the reaction, respectively.

3. Computer-Derived Parameters

Though specific first-order rate constants are useful, they nevertheless require replotting the data in the event a linear log recorder is not available and require moreover calculation of the rate constants. By means of a Digital Equipment Corporation PDR-15 computer equipped with a Tetronix T-4002 graphic computer terminal and hard copy printer, it was possible to supply kinetic data (absorbance as a function of time) directly into the computer and obtain computer-derived constants. The computer program is written in Focal language (Copyright, Digital Equipment Corp.; cf. Martin et al., 1973). The input data were presented visually and if satisfactory a curve was plotted and, if desired, photostatic copy could be obtained. The program was designed to calculate four parameters [Eqn.(1)].

$$Y = D/[1 + \mathrm{Exp}(X-B)/C] + E \qquad (1)$$

Here, Y is equal to absorbance values, X is equal to time expressed in seconds, D is a spread factor, i.e., $A_i - A_\infty$, where A_i is the initial absorbance observed during the prolytic period (= 1.0 absorbance units) and A is the final absorbance, C is equal to the slope of the lytic phase, B is equal to the midpoint time value, i.e., the value of t at which A_t is equal to D, and E is the off-set constant and is approximately equal to A_∞, as will be demonstrated. In relation to the other rate parameters, C^{-1} is directly related to pseudo first-order rate constant, k_ψ, and B is closely related to the length of the prolytic period, t_p.

IV. RESULTS AND DISCUSSION

One of the very useful methods of characterizing the action of marine biotoxins is the effect on red blood cells. The events that precede and accompany toxin-induced hemolysis are thus very interesting, though in the past problems have arisen because of limitations of analysis or methods. Ponder (1948) noted that in several instances a truly theoretical treatment had failed because one parameter or another could not be directly evaluated.

During the years since the publication of his monograph, several useful experimental approaches and analytical techniques have become available that mitigate the analytical problems to a considerable extent. Spectrophotometric methods, for example, permit more adequate characterization of rates of toxin binding and lysis. Radiolabeled toxins are now available that can be used to indicate the extent of binding, and these results compare favorably with those results of binding obtained through use of a flow microcalorimeter (Binford et al., 1973). Finally, with the aid of data acquisition systems and computers, useful mathematical treatments can be used to indicate the mode of action of the marine biotoxin. As we will discuss briefly in the following sections, the combination of these approaches and techniques bring us closer to a kinetic if not an operational description of toxin-membrane interaction.

A. Kinetic Studies

The kinetic patterns of prymnesin-induced hemolysis of red blood cells suggested a simple model which showed a reasonable analogy with enzyme kinetics, particularly with respect to the Michaelis-Menten or Henri treatment (Martin and Padilla, 1971a). We recognize that this must be an oversimplification, particularly in view of the known complexity of both prymnesin and the cell membrane, but we suggest the analogy may be valid for several reasons.

First, the two parameters used to calculate values of k_ψ from the usual Michaelis-Menten kinetics [Eq. (2)] adequately describe the lysis step:

$$k_\psi = \frac{k_{\psi}m\ (L,\ HD_{50})}{K_m + (L,\ HD_{50})} \qquad (2)$$

Here, the terms have essentially the usual meanings: $k_{\psi m}$ is the maximum observed pseudo first-order rate constant, K_m is a Michaelis constant, and L, HD_{50} is the concentration of prymnesin, expressed in hemolytic units.

Second, the agreement between observed and calculated [Eq. (2)] values of k_ψ was good in the low and intermediate concentration ranges of prymnesin (HD_{50} = 0.125–25). The agreement did fail at high concentration ranges (HD_{50} = 25–1250), however, and the calculated values are higher than observed (Martin and Padilla, 1971a). It appears that the disparity could be ascribed to a saturation effect, to inhibition by the hemolysin, or to inhibition by by-products.

Third, the complexity of the process(es) associated with k_ψ is indicated by the variation of rate constants with temperature. A conventional linear Arrhenius relationship was obtained for the first specific rate constant, k´, but not for k_ψ, and use of a modified Arrhenius plot $(k/T^{1/2})$ vs. 1/T did not produce a linear plot. The absence of linearity suggests the complexity of the reaction.

B. Inhibition Studies

Since our data did not permit a distinction between the possibility that the toxin (L) interacts with a single or several active complexes (C) at suitable receptor sites on the erythrocyte membrane (E), we extended the analysis to the interaction between hemolysin, erythrocytes, and inhibitor (I) [cf. Eq. (3).]

$$E + L \underset{k_{-1}}{\overset{k_1}{\rightleftarrows}} C \underset{k_{-2}}{\overset{k_2}{\rightleftarrows}} E + P \qquad (3a)$$

$$L + I \underset{k_{-3}}{\overset{k_3}{\rightleftarrows}} L_i \qquad (3b)$$

$$E + L_i \underset{k_{-4}}{\overset{k_4}{\rightleftarrows}} C_i \qquad (3c)$$

Specific rate constants were determined as a function of concentration of inhibitor lipids added to an erythrocyte suspension containing 12.5 HD_{50} of prymnesin (intermediate range). Effects of inhibition were expressed in two ways: first, as relative rate constants, k_{rel} (the ratio of the constants in the presence and absence of inhibitor); and second, as the inhibition parameter, IP $= [(1-k_{rel})/k_{rel}]$. The results indicated that cephalin was a more active inhibitor on a molar basis than cholesterol at concentrations at which cephalin was not a hemolysin (0.01-0.1 mM, cf. Rauckman and Padilla, 1970).

The mechanism of inhibition may be inferred from a diagnostic plot, assuming that the simplistic model represented by Eq.(3) is operative. Inhibition of lysis could occur via three major pathways: (1) decrease in effective concentration of hemolysin through formation of a toxin-inhibitor complex [L_i, Eq. (3b)]; (2) membrane-complex interaction (L_i) in competition with hemolysin [Eq. (3c) vs. (3a)]; and (3) competitive displacement of the hemolysin by the inhibitor. The three pathways should be distinguished by the diagnostic plot. Inhibitor concentration as a function of inhibition parameter (Fig.3) for pathway 1, should result in a hyperbola (concave downward); for pathway 2, a linear relationship is expected; for pathway 3, a curve (concave upward) is expected, according to Reiner (1969). Pathway 1 seemed to apply with both cephalin and cholesterol in the range studied, pathway 2 was found to hold, however, in the interaction between prymnesin and Gymnodinium breve toxin. We can anticipate further studies of inhibitors of toxin-membrane interaction will be rewarding.

The analytical approaches described above cannot be used to indicate the stoichiometry of the toxin-membrane interaction. A Hill-type plot, however, can be used to indicate the stoichiometry of inhibition, according to Loftfield and Eigner (1969). For the simplistic enzymelike model, the data should be ammeniable to a Hill-type plot treatment in which the logarithm of a function of degree of inhibition shows a linear relationship with the logarithm of the inhibitor concentration. This type of analysis has been neglected for toxin-erythrocyte systems, but the approach could be a useful one because in many instances the slope of the straight line of the Hill-type plot has an integral value. The value systems for enzyme indicates the integral value of the slope indicates that one, two, or more (depending upon the value) inhibitor molecules react cooperatively to inactivate or inhibit the enzyme; a nonintegral value is visualized as a consequence of incomplete inhibition (Loftfield and Eigner, 1969). The approach may be applied

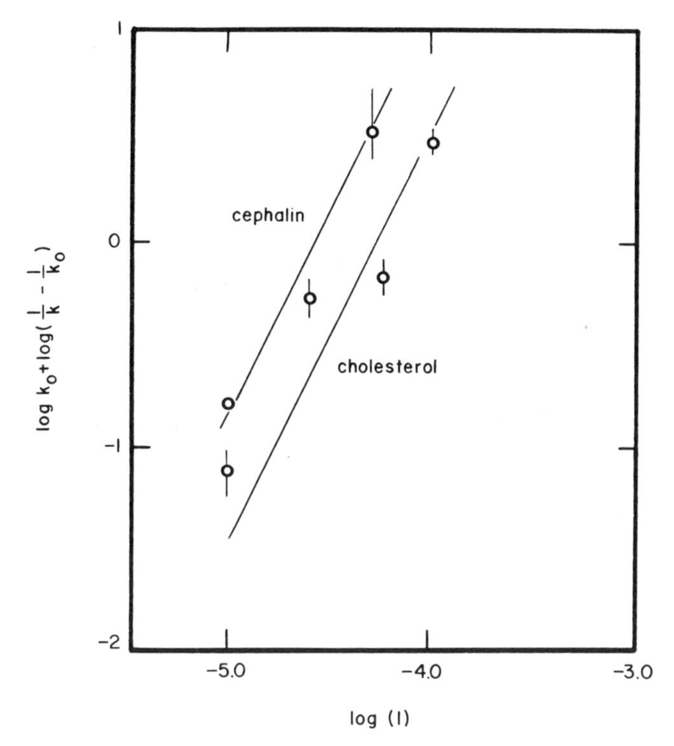

FIG. 4. Hill-type plot for lipid inhibition of prymnesin-induced hemolysis of rabbit erythrocytes using the data of Martin and Padilla (1971a). Straight lines of slope 2.0 are indicated.

to the lipid-prymnesin-erythrocyte system described above, and the Hill-type plot (Fig.4) has a linear relationship with a slope of 2. The other diagnostic plot indicated a formation of a toxin-inhibitor complex, and the second diagnostic plot would suggest that two lipid molecules (cephalin or cholesterol) are associated with each effective unit of prymnesin in the activated complex. The approach is a useful one, though additional information would be welcome through binding studies.

C. Toxin Binding

The accumulation of toxin at the erythrocyte surface during the prolytic phase is a significant feature of hemolysis (Martin and Padilla, 1971a; Binford et al., 1973). About 40% of labeled toxin was bound to rabbit erythrocyte membrane and about 10% was loosely bound, as indicated by the ease of removal with methanolic washing. Results with bovine blood are similar (Binford et al., 1972). The fraction of prymnesin bound appeared to be independent of the number of cells added (over the range 9-900 million cells/0.5 ml).

No fundamental difference in the binding of prymnesin to red cells or to osmotically lysed cells or ghosts was observed (Martin and Padilla, 1971a). The binding appeared to be unaffected by deformation of membrane structure that must have occurred during lysis.

Binding tendencies are, however, a function of pH of the solution. Radiochemical studies with [^{3}H]-labeled toxin show the maximum binding of prymnesin to bovine red cells occurs in the pH range 4.5-5.5, and that the binding tendency decreases in the range 5.5-7. These observations can be correlated to a high degree with microcalorimetric measurements using a flow microcalorimeter (Binford et al., 1973). These experiments provide useful information, but additional studies will be necessary to provide a better description of the receptor sites on the red cells.

D. Other Techniques

The effect of prymnesin and *G. breve* toxin on the permeability of red cells has been examined. The effect of the first toxin on potassium fluxes using red cells cannot be measured because of the destructive action of the toxin on red cell membranes (Dafni and Giberman, 1972). One *G. breve* toxin did not cause hemolysis (Martin et al., 1972) and does appear to affect potassium transport according to our preliminary results. The relative fluxes (potassium fluxes in the presence and absence of *G. breve* toxin) depend on *G. breve* concentration, and for a given concentration the logarithm of the relative flux is a linear function of external potassium concentration in a normal Ringers buffer for several erythrocytes (human, rabbit, and bovine). It is interesting to consider whether the relative flux is also a function of internal concentration as well. It is possible to test this by means of genotype sheep blood using HK (high concentrations of potassium and low concentrations of sodium) and LK (low concentrations of potassium and high concentrations of sodium) sheep blood, for which the two cell types are morphologically indestinguishable. (Kerr, 1937; Hoffman, 1969). We have found the logarithm of potassium fluxes for the HK and LK blood is independent of external concentration and that the relative fluxes are independent of blood type.

V. CONCLUDING REMARKS

We have attempted to suggest some of the features of prymnesin-membrane interactions, together with some of the less common procedures that may be useful in characterizing the basic physiological activity of marine biotoxins. Admittedly, these are at the early stage of bringing into focus those parameters that will, in due time, permit a quantitative and sequential description of the physiochemical events that are concomittant with hemolysis. Unfortunately, it is presently impossible to determine how prymnesin affects cell lysis, and at best we can reach only a few general conclusions, particularly in view of the uncertainty concerning the molecular architecture of the toxin (Paster, 1968; Ulitzur and Shilo, 1970b).

It seems likely that an approach that we have used briefly - flow microcalorimetry - may assume greater significance in our efforts to understand the action of prymnesin.

ACKNOWLEDGMENTS

This research was supported by Grant No. SD 00120 from the Food and Drug Administration, Consumer Protection and Environmental Health Service, U.S. Public Health Service. One of us (D.F.M.) gratefully acknowledges a Career Development Award (K04-GM 42569-03, National Institute of General Medical Sciences) from the Public Health Service.

REFERENCES

Abbott, B.C., and Ballantine, D., J. Marine Biol. Assoc. U.K. 36, 169 (1957).

Bergmann, F., and Kidron, M., J. Gen. Microbiol. 44, 233 (1966).

Bergmann, F., Parnas, I., and Reich, K., Brit. J. Pharmacol. 22, 47 (1964).

Binford, J.S., Jr., Martin, D.F., and Padilla, G.M., Biochim. Biophys. Acta 291, 156 (1973).

Burke, J.M., Marchisotto, J., McLaughlin, J.J.A., and Provasoli, L., Ann. N.Y. Acad. Sci. 90, 837 (1960).

Collier, Albert, "Pütter in Retrospect," paper presented at the Symposium on Physiologically Active Compounds from Marine Organisms, Univ. of South Florida, St. Petersburg, Florida, November 1971 (unpublished).

Connell, C.H., and Cross, J.B., Science 112, 359 (1950).
Cummins, J.M., Jones, A.C., and Stevens, A.C., Trans. Amer. Fish. Soc. 100(1), 112 (1971).
Dafni, Z., J. Protozool. (Suppl.) 16, 38 (1969).
Dafni, Z., and Giberman, E., Biochim. Biophys. Acta 255, 380 (1972).
Dafni, Z., and Shilo, M., J. Cell Biol. 28, 461 (1966).
Halstead, B.W., "Poisonous and Venomous Marine Animals of the World," Vol. I. U.S. Government Printing Office, Washington, D.C., 1965.
Hoffman, P.G., Jr., Doctoral dissertation, Duke University, Durham, North Carolina, 1969.
Hutton, R.F., Quart. J. Fla. Acad. Sci. 19, 124 (1956).
Ingle, R.M., Univ. Miami Marine. Lab. Spec. Serv. Bull. 9. 1 (1954).
Kerr, S.E., J. Biol. Chem. 117, 227 (1937).
Loftfield, R.B., and Eigner, E.A., Science 164, 305 (1969).
McFarren, E.F., Tanabe, H., Silva, F.J., Wilson, W.B., Campbell, J.E., and Lewis, K.H., Toxicon 3, 111 (1965).
McLaughlin, J.J.A., J. Roy. Microscop. Soc. 83, 317 (1958).
Martin, D.F., and Chatterjee, A.B., Fish. Bull. 68, 433 (1970).
Martin, D.F., and Padilla, G.M., Biochim. Biophys. Acta 241, 213 (1971a).
Martin, D.F., and Padilla, G.M., Environmental Letters 1, 199 (1971b).
Martin, D.F., Padilla, G.M., Heyl, M.G., and Brown, P.A., Toxicon 10, 285 (1972).
Martin, D.F., Padilla, G.M., and Dessent, T.A., Anal. Biochem. 51, 32 (1973).
Padilla, G.M., J. Protozool. 17, 456 (1970).
Parnas, I., Israel J. Zool. 12, 15 (1963).
Parnas, I., Abbott, B.C., Toxicon 3, 133 (1965).
Parnas, I., and Spiegelstein, M.Y., Fish. and Fish Breeding Israel. 1, 13 (1963).
Paster, Z., Rev. Intern. Oceanogr. Med. 10, 249 (1968).
Paster, Z., and Abbott, B.C., Toxicon 7, 245 (1969).
Ponder, E., "Hemolysis and Related Phenomena," p.398. Grune and Stratton, New York, 1948.
Prakash, A., J. Fish. Res. Bd. Canada 24, 1589 (1967).
Prakash, A., and Taylor, F.J.R., J. Fish. Res. Bd. Canada 23, 1265 (1966).
Rauckman, B., and Padilla, G.M., 14th Ann. Meeting Biophys. Soc., Baltimore, Md., Feb. 27-29, 1970.

Ray, S.M., and Wilson, W.B., U.S. Fish. Wildl. Serv. Fishery Bull. 123, 469 (1957).

Reich, K., Bergmann, F., and Kidron, M., Toxicon 3, 33 (1965).

Reiner, J.M., "Behavior of Enzyme Systems," 2nd ed. Van Nostrand Reinhold, New York, 1969.

Rounsefell, G.A., and Nelson, W.R., U.S. Fish Wildl. Serv. Spec. Sci. Rep. Fish. No. 535 (1966).

Ryther, J.H., in "The Luminescence of Biological Systems" (F.H. Johnson, ed), pp.387-414. Amer. Assoc. Advan. Sci., 1955.

Schantz, E.J., J. Med. Pharmac. Chem. 4, 459 (1961).

Schantz, E.J., in "Properties and Products of Algae" (J.E. Zajic, ed.), pp. 83-96. Plenum Press, New York, 1970.

Shilo, M., and Shilo, M., Appl. Microbiol. 1, 330 (1953).

Shilo, M., Bacteriol. Rev. 31, 180 (1967).

Shilo, M., in "Microbial Toxins" (S. Kadis, A. Ciegler, and S. J. Ajl, eds.), Vol. 17, pp. 67-103. Academic Press, New York, 1971.

Shilo, M., and Rosenberger, R.F., Ann. N.Y. Acad. Sci. 90, 866 (1960).

Steidinger, K.A., Burklew, M.A., and Ingle, R.M., in "Marine Pharmacognosy" (D.F. Martin and G.M. Padilla, eds.), Chapt. 6. Academic Press, New York, 1973.

Trief, N.M., Spikes, J.J., Ray, S.M., and Nash, J.B., in "Toxins of Plant and Animal Origin" (A. DeVries and E. Kochva, eds.) pp. 557-577. Proc. Second Internat'l Sympos. on Plant and Animal Toxins, Tel Aviv, Israel, 1970.

Ulitzur, S., and Shilo, M., J. Gen. Microbiol. 36, 161 (1964).

Ulitzur, S., and Shilo, M., J. Protozool. 13, 332 (1966).

Ulitzur, S., and Shilo, M., J. Gen. Microbiol. 62, 363 (1970a).

Ulitzur, S., and Shilo, M., Biochim. Biophys. Acta 201, 350 (1970b).

Wiberg, G.S., and Stephenson, N.R., Toxicol. Appl. Pharmacol. 2, 607 (1960).

Woodcock, A.H., J. Marine Res. 7, 56 (1948).

Wilson, W.B., Doctoral dissertation, Texas A & M University, College Station, Texas, 1965.

Yariv, J., and Hestrin, S., J. Gen. Microbiol. 24, 165 (1961).

Chapter 9

PARTIAL PURIFICATION AND BIOLOGICAL PROPERTIES OF AN EXTRACT OF THE GREEN SPONGE, HALICLONA VIRIDIS

Robert E. Middlebrook* and Carl H. Snyder
Department of Chemistry, University of Miami
Coral Gables, Florida
and
Arsenio Rodriguez Mercado and Charles E. Lane
Rosenstiel School of Marine and Atmospheric Science
University of Miami
Miami, Florida

I. INTRODUCTION

Our interest in the chemistry of the green sponge, *Haliclona viridis*, was stimulated by the disclosure of several workers that alcoholic or aqueous extracts of this animal show a variety of biological activities. For example, Jakowska and Nigrelli (1960) have commented on the antimicrobial and lethal properties of *H. viridis* extract as well as its hypotensive effect in mammals (Nigrelli et al., 1961). Sigel et al. (1970) have observed the cytotoxic effects of crude sponge extract against KB human cancer cells. The name "halitoxin' has been proposed by Baslow and Turlapaty (1969) to designate a water-soluble toxin extracted from oven-dried *H. viridis*. They have also established toxicity levels for this substance and have demonstrated its *in vivo* antitumor activity.

We have confirmed several of these observations and have established the presence of *in vitro* anticholinesterase and direct hemolytic activities in partially purified aqueous extracts of the green sponge. In addition, we have become increasingly susceptible to certain effects presumably exerted by *Haliclona viridis* extract. Symptoms include running eyes, running nose, sneezing, and coughing.

* Present address: Department of Marine Sciences, University of Puerto Rico, Mayaquez, Puerto Rico.

Our immediate goals in examining the chemistry of _H. viridis_ are in line with our general interest in marine toxins. One of the objectives is the purification of the lethal component of the sponge extract. Of equal importance is an attempt to differentiate lethal and antineoplastic activities, and if preliminary studies suggest they are chemically distinguishable, to isolate and characterize each structurally. A third principal concern is an attempt to determine the extent of overlap of biological activities among the various purified fractions.

II. MATERIALS AND METHODS

Sponges were normally collected at the north end of Key Biscayne in the immediate Miami area, although they are abundant in several shallow water locales throughout southern Florida. Animals were freed of detritus and soaked in cold phosphate buffer, pH 6, for 12-18 hr. The viscous dark green extract was decanted, clarified by paper filtration, and exhaustively dialyzed against distilled water at $4^{o}C$ for 48 hr. All activity was present in the dialysis retentate. The procedure was repeated three times with each batch of sponges with the buffer replenished immediately following decantation of the extract. Normally, 6-8 kg wet weight of sponges afforded about 25 mg of dry crude extract.

Ion-exchange chromatography on carboxymethyl cellulose (Whatman CM-52) was selected as the initial method of purification. Stepwise elution of products from the column with aqueous sodium chloride was ultimately judged more satisfactory than a gradient elution process. In a typical run, 1500-2000 ml of crude dialyzed sponge extract was filtered through 200 gm of carboxymethyl cellulose in a 16 cm diameter Büchner funnel. All activity was retained by the exchanger, and a quantity of dark green material separated in the filtrate. Elution of products was effected by exhaustive washing with 0.1 to 1.0 M aqueous sodium chloride with the ionic strength of each 200 ml wash increased by 0.1 M increments. The filtrate was recycled at each step two or three times to produce an essentially colorless effluent.

Gel filtration was accomplished on 2 x 85 cm Sephadex G-100 columns with distilled water used as eluent. Column effluents were monitored at 260 nm with an LKB Uvicord.

LD_{50}'s were estimated by the up and down method of Dixon (1965) for small samples. Doses were administered IV to Swiss white mice in 0.1 ml of physiological saline. Cytotoxicity was determined in tissue culture against KB human cancer cells and is expressed as micrograms per milliliter required to effect 50% inhibition of cell growth. *In vitro* anticholinesterase activity was measured with a modification of Sigma Chemical Company's cholinesterase kit for the colorimetric determination of cholinesterase. The concentration of sponge fractions tested was 3.5 μg/ml. Results of hemolytic studies, carried out with fresh human blood, are expressed as percent hemolysis relative to a distilled water control (100% hemolysis). Test solutions were 10 μg/ml.

Fraction complexity was evaluated by disc gel electrophoresis on a Buchler Polyanalyst. Running time was 3 hr at 2-3 mA/tube with migration toward the cathode. The separation gel was 7.5% acrylamide, and the buffer was diethylbarbituric acid/tris, pH 7.5, Williams and Reisfeld (1964). Gels were stained overnight in 0.25% amido black and destained with 7% aqueous acetic acid.

III. RESULTS

A LD_{50} of 1750 μg/kg was established for the crude dialyzed extract of *H. viridis*. Chromatography of the product on carboxymethyl cellulose with gradient elution from 0 to 1 M sodium chloride and monitoring at 260 nm produced a multiplicity of effluent peaks. A repetition of this chromatography employing stepwise elution of the column dislodged a quantity of pigment at each ionic strength of sodium chloride. Colors varied from blue to green, dark green, and finally brownish black. Lethal activity was present in all fractions except the first (0.1 M sodium chloride) at dose levels from 10-20 μg/mouse. LD_{50} levels were typically 750 μg/kg. Effects in the mouse, evident within 5 min, included swollen and running eyes, disorientation, respiratory distress, and paralysis. Death occurred within 1 hr.

Antineoplastic activity was not differentiated from lethal activity and showed a ED_{50} of 5 μg/ml. Hemolytic and anticholinesterase activities showed a gradual rise with increasing ionic strength of the eluent (Fig.1).

There was no clear differentiation of activities on carboxymethyl cellulose with the exception of the first two lethal fractions which have no anticholinesterase or hemolytic activities. Fractions were freed of salt by dialysis and lyophilized. All were amorphous and varied in color from tan to

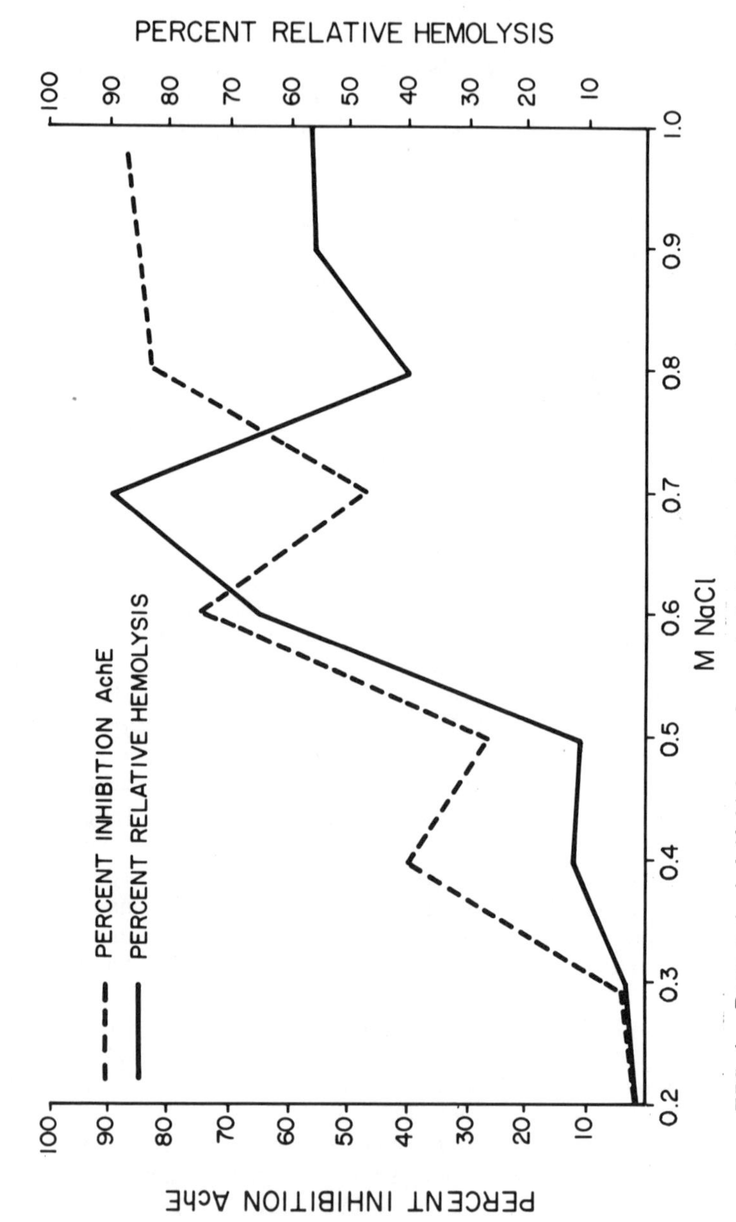

FIG.1. Percent inhibition of acetylcholinesterase and percent relative hemolysis for fractions eluted from carboxymethyl cellulose.

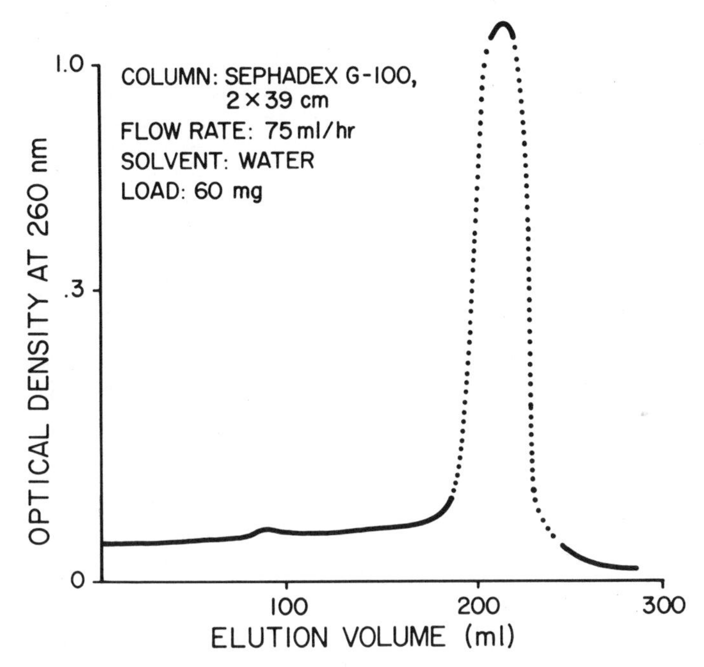

FIG. 2. Sephadex G-100 gel filtration.

brownish black. Fractions eluted at high sodium chloride molarities (0.7-1 M) were quite hygroscopic. Disc gel electrophoresis of all fractions produced two bands, the lower being wider and much more heavily stained. Gels typically displayed a stairstep pattern of this band, reflecting the increased ionic strength required to dislodge each fraction from the cation exchanger. Highly positively charged substances migrated more strongly toward the cathode.

It was decided to treat each fraction from carboxymethyl cellulose separately in further purification studies, since combining lethal fractions, for example, would require the pooling of nearly all fractions. In addition, a clear separation of pigments and an increase in lethality indicated that some purification was being achieved even though activities were spread over a considerable fractionation range.

Preliminary trials had indicated that gel filtration on Sephadex G-100 was a satisfactory step subsequent to cation-exchange chromatography. Fractions from carboxymethyl cellulose behaved similarly on G-100 with activities centered in two partly resolved colored bands emerging from the column at approximately twice the void volume of the column. The elution pattern is shown in Fig.2. This figure is somewhat deceptive since it is known that inactive material not absorbing at 260 nm emerges from the column at the void volume and continues until absorbance (activity) begins. The peak in Fig.2, though reasonably symmetrical, is almost certainly a mixture of products having the same approximate size and/or shape. No activity was present in material on either side of the peak. Lethal, antineoplastic, hemolytic, and anticholinesterase activities were distributed in the two colored bands (green preceding yellow) emerging from the column. Attempts to resolve these completely on G-100 and other Sephadex sizes were unsuccessful.

Lethal activity for lyophilized fractions was improved (LD_{50} 220 μg/kg). but antineoplastic activity remained unchanged (ED_{50} 4.1 μg/ml). Hemolytic and anticholinesterase activities were marginally increased for those fractions eluted from carboxymethyl cellulose by low ionic strengths of sodium chloride. In general, recovery of product from G-100 was good, but a slight amount of pigmented material was invariably retained and probably is highly positively charged.

Surprisingly, it was observed that when products from G-100 were exhaustively dialyzed against distilled water, activity was recovered in both retentate and dialysate, although in previous purification steps all activity had been in retentates alone. The retentate showed a LD_{50} of 175 μg/kg, heightened only marginally over the gel filtration product. Lethality of the dialysate was reduced over threefold to a LD_{50} of 750 ug/kg. Interestingly, antineoplastic activities were nearly identical for both retentate and dialysate showing ED_{50}'s of 4.8 and 5.1 μg/ml, respectively. Lethality may be halved by heating both fractions at 60°C for 1/2 hr, but even at 100°C it is not completely destroyed after 1/2 hr.

Anticholinesterase and hemolytic activities were nearly equal in the retentate and dialysate when precursor material had been eluted from carboxymethyl cellulose at relatively high ionic strengths of sodium chloride (0.6-1.0 M). In contrast for those precursor fractions eluted at lower concentrations of sodium chloride (0.2-0.4 M), both anticholinesterase and hemolytic activities were present mostly in the retentate with only minimal activities appearing in the dialysate. Disc gel electrophoresis of 100 μg each of these fractions showed a wide major band appearing in both fractions and contrasting only in intensity. Positive Dragendorffs, ninhydrin, and Lowry tests are obtained with all partially purified fractions. In addition, fractions spotted on cellulose may be visualized with ultraviolet light.

IV. DISCUSSION

In line with our stated goals, we have been successful in partially purifying the lethal component of *H. viridis* extract tenfold. Our purest fractions, however, are amorphous and very slightly pigmented, suggesting the need for still further purification. We have not as yet carried out any hydrolytic experiments, having preferred instead to pursue acquisition of purer material. Attempts to evaluate the complexity of this material by thin-layer chromatography have been unsuccessful. Silica gel, microcrystalline cellulose, and carboxymethyl cellulose plates have been employed with many different solvent systems. Spotted material did not move or gave only streaking from the origin. Thus we have relied on disc gel electrophoresis for evaluation of complexity. A second visualization system is still being sought.

Our second goal, differentiation of lethal and antineoplastic activities, remains to be achieved. In our purest fractions, cytotoxicities are identical, ED_{50} of 5 μg/ml. However, these same fractions display quite different lethalities, LD_{50}'s of 175 μg/kg and 750 μg/kg. These facts suggest that lethality and cytotoxicity have been differentiated to some degree. This conclusion is strengthened by the observation that lethality can be halved by heating the purest fractions without affecting anticholinesterase and hemolytic activities. Preliminary evidence suggests cytotoxicity is also undiminished by heating. However, final clarification of this question must await further purification accompanied by *in vitro* and *in vivo* testing of the product.

With regard to our third concern, it is obvious that in the present state of purification there is indeed considerable overlap of activities. It is tentatively concluded that in *H. viridis* extract there is present a series of compounds, closely related in size and differing in charge, which manifest to a greater or lesser degree all of the activities described. The chemical nature of these may only be speculated on at present.

ACKNOWLEDGMENTS

R.E. Middlebrook and C.H. Snyder wish to acknowledge support of these studies by the American Cancer Society, Florida Division, Inc., through Grant No. F 71 UM-2. The same authors take pleasure in acknowledging their indebtedness to Bill Lichter and Larry Wellham of the laboratory of Dr. M.M. Sigel, Department of Microbiology, University of Miami, who carried out cytotoxicity tests.

REFERENCES

Baslow, M.H., and Turlapaty, P., Proc. West. Pharmacol. Soc. 12, 6 (1969).

Dixon, W.J., J. Amer. Statist. Ass. 60, 967 (1965).

Jakowska, S., and Nigrelli, R.F., Ann. N.Y. Acad. Sci. 90, 913 (1960).

Nigrelli, R.F., Baslow, M.H., and Jakowska, S., Amer. Soc. Microbiol. p. 83 (1961).

Sigel, M.M., Wellham, L.L., Lichter, W., Dudeck, L.E., Gargus, J.L., and Lucas, A.H., Marine Tech. Soc. p. 281 (1970).

Williams, D.E., and Reisfeld, R.A., Ann. N.Y. Acad. Sci. 121, 373 (1964).

Chapter 10

PHYSIOLOGICALLY ACTIVE SUBSTANCES FROM ECHINODERMS

George D. Ruggieri and Ross F. Nigrelli

Osborn Laboratories of Marine Sciences
New York Aquarium, New York Zoological Society
Brooklyn, New York

Twenty years ago, Nigrelli (1952) in New York and Yamanouchi (1955) in Japan independently coined the term holothurin to signify the toxic material in sea cucumbers. Yamanouchi applied the term to the toxic extract from the body wall of *Holothuria vagabunda* Selenka and other Southeast Asian holothuroids, and Nigrelli applied it to the toxic material in the Cuverian tubules of the Bahamian sea cucumber, *Actinopyga agassizi* Selenka. In the ensuing years, extensive research on these and related members of the phylum Echinodermata had progressed along two levels. Certain studies (cf. Chanley et al., 1966; Friess et al., 1967) are concerned with the chemical purification (Holothurin A) and modification (desulfated holothurin A) and biological activities of holothurin. Other studies are being directed at uncovering similar biologically active substances from other classes of the echinoderms.

Early studies by Nigrelli and associates dealt with the action of crude Holothurin (sun-dried, powdered, Cuverian organs of *Actinopyga agassizi* Selenka) in suppressing the growth of Sarcoma-180 and Krebs-2 ascites tumors in Swiss mice (Nigrelli, 1952; Nigrelli and Zahl, 1952; Sullivan et al., 1955; Sullivan and Nigrelli, 1956). Other studies focused on the highly toxic quality of holothurin on a variety of test organisms. This substance suppressed the growth of watercress root hairs and caused necrosis in onion root tips (Nigrelli and Jakowska, 1960); inhibited the growth of a variety of protozoans (Nigrelli and Zahl, 1952; Nigrelli and Jakowska, 1960);

interfered with regeneration in planaria (Quaglio et al., 1957); retarded pupation in fruit flies (Goldsmith et al., 1958); modified developmental patterns in echinoderm eggs (Ruggieri and Nigrelli, 1960); and was toxic to a number of fish species (Nigrelli, 1952; Nigrelli and Jakowska, 1960; Rio et al., 1965).

Chanley and co-workers (Chanley et al., 1966; Chanley and Rossi, 1969a; Chanley and Rossi, 1969b) have proposed a provitional structure for Holothurin A. It contains a complex steroidal nucleus, to which are attached the monosaccharides D-glucose, D-xylose, D-quinovose, and 3-0-methylglucose and a half-esterified sulfuric acid moiety. It is this sulfuric acid moiety that gives the saponin molecule its anionic character. Chanley and Rossi (1969b) have reported that the sulfate moiety is attached to the xylose. A neutral desulfated derivative has been obtained by chemical removal of the sulfate charge center from anionic holothurin A.

Friess et al. (1965, 1967, 1968, 1970) have highlighted the important role played by the negative-charge site in Holothurin A in effecting irreversible blockade of cholinergic neuromuscular transmission and in evoking direct contractural response from striated muscle. Desulfated holothurin A is less potent and less irreversible in its blockade of nerve node excitability or neuromuscular transmission than that of the holothurin A (Friess et al., 1970). Friess and co-workers have shown further that treatment with desulfated holothurin A affords protection against the irreversible effects of anionic holothurin A on peripheral tissues, such as the node of Ranvier in frog sciatic nerve fibers (Friess et al., 1968) and the *in vitro* phrenic nerve-diaphragm preparation from the rat (Friess et al., 1967). Recent studies (Friess et al., 1970) have shown that the action of anionic holothurin A is one order of magnitude more potent than the neutral desulfated Holothurin A in destroying ganglionic excitability. Desulfated Holothurin A, however, protected ganglionic excitability against the destructive action of anionic holothurin A when used at levels below its own effective blockade concentrations (Friess et al.., 1970). However, desulfated holothurin A does not protect central nervous system receptors from the destructive action of anionic holothurin A (Friess et al., 1968).

Lasley and Nigrelli (1970) studied the effects of crude Holothurin on the physical and chemical behavior of human phagocytes during engulfment of staphylococci. They noted that the phagocytic activity of leukocytes increased in the presence of progressively higher levels of holothurin. This activity appears to involve effects on the surface membranes and alterations of intracellular granules with a subsequent liberation of lytic enzymes. Lasley and Nigrelli (1971) have also reported on the effects of crude holothurin, holothurin A, and desulfated holothurin A on ameboid movement of leukocytes. Weak concentrations of these substances stimulated leukocyte migration while higher concentrations inhibited leukocyte migration. The sequence of effectiveness of these substances in stimulating and inhibiting leukocyte migration was: anionic Holothurin A > crude holothurin > desulfated holothurin A. It is interesting to note that concentrations of desulfated holothurin A needed to stimulate leukocyte migration were 50 times greater than that of anionic holothurin A. In order to inhibit leukocyte migration, the effective concentration of desulfated Holothurin A was 3000 times that of Holothurin A.

Styles (1970) has shown that crude Holothurin administered to rats prior to or simultaneous with the inoculation of trypanosomes resulted in diminished parasitemia. A slight increase in parasite load was noted in rats treated after trypanosome infection. Rats treated with 100 μg of crude holothurin on 5 successive days and then challenged with 7×10^3 trypanosomes 3 days later had an average of 3800 trypanosomes/mm^3 of blood at the peak of infection as compared with 17,500 organisms/mm^3 of blood in the controls. Rats injected with 250 μg of crude holothurin and 7×10^3 trypanosomes at the same time had 7300 trypanosomes/mm^3 of blood compared to 16,700 trypanosomes/mm^3 of blood in the controls. Rats treated with crude holothurin after inoculation of trypanosomes had approximately one-and-a-half times as many trypanosomes as the control. The author has suggested that since immunity to trypanosomes is mediated through cellular and humoral mechanisms, both of which are closely associated with the reticuloendothelial system, stimulation or damage to either or both of these mechanisms may account for the inhibitory and stimulatory effect. Recent experiments (Styles and Hoshino, 1971) indicate that holothurin A inhibits trypanosome infectivity only when administered simultaneously. Holothurin A, therefore, unlike crude Holothurin, is not effective when administered prior to infection.

The effects of holothurin A on the automaticity, action potential characteristics and associated conduction phenomena of rabbit sinus node and atrioventricular nodal cells and dog Purkinje fibers have been investigated recently (Ricciutti and Damato, 1971). The rate of sinus node automaticity fell from 72 beats/min and an action potential height of 80 mV in the control to 48 beats/min and an action potential height of 43 mV when rabbit sinus node cells were treated with 10 μg/ml of holothurin A for 20 min. Thirty-minute perfusion with 10 μg of holothurin A followed by a 20 min wash period with Tyrode solution resulted in a return to 72 beats/min, but the action potential height dropped to less than 20 mV. A concentration of 10 μg/ml of holothurin A for 15 min caused a 40-50% increase in conduction time through the atrioventricular node. Holothurin A also produced a 50% decrease in the automatic rate of spontaneously beating Purkinje cells. These studies show that both excitation and conduction phenomena of myocardial cells can be modified by Holothurin A. The ability of this substance to suppress sinus node automaticity suggests a possible use in sinus node arrythmias and tachycardias.

The present report deals with certain physiological activities exhibited by crude extracts of a number of other species of echinoderms. The following new species, all collected from the West Indies except the Crown-of-Thorns sea star, *Acanthaster planci*, obtained from the Indo-Pacific, were investigated:

I. Holothuroidea:

Holothuria parvula (Selenka)
Holothuria arenicola Semper
Holothuria mexicana Ludwig
Holothuria densipedes H.L. Clark
Astichopus multifidus (Sluiter)

II. Asteroidea:

Acanthaster planci (Linnaeus)

III. Echinoidea:

Diadema antillarum Philippi
Mellita quinquiesperforata (Leske)
Clypeaster rosaceus (Linnaeus)
Meoma ventricosa (Lamarck)

IV. Ophiuroidea:

Ophiocoma echinata (Lamarck)

V. Crinoidea:

Nemaster rubiginosa (Pourtales)

All specimens were extracted according to the procedure used by Rio et al., (1965). The lyophilized material was then tested on the following organisms:

1. Common killifish, Fundulus heteroclitus (Linnaeus), for ichthyotoxic properties

2. Gametes of the sea urchin, Arbacia punctulata (Lamarck) for effects on sperm motility, eggs, and developing embryos

3. KB oral carcinoma cells and Gray Seal Kidney (GSK) cells for effects on cells in vitro

4. Candida albicans and Saccharomyces cerevisiae for antifungal properties

I. EFFECTS ON FUNDULUS HETEROCLITUS

Killifish are hardy and easily maintained under laboratory conditions. Eggs were manually stripped from ripe females and fertilized with sperm from sacrificed males. The young fish were used between 1 and 2 weeks after hatching and measured 8-9 mm in length. These fish have proved to be very suitable for toxicity studies. They are parasite-free and afford a clear view of the heart, since pigment deposition has not yet progressed to the point of obscuring the heart beat. Cessation of heart beat, then, serves as the end point rather than the stilling of opercular movement as was the case with adult killifish.

All the sea cucumber extracts were lethal to killifish in concentrations of 2 μg/ml for time periods ranging from 120 to 145 min. Anionic holothurin A was the most toxic, killing the fish in 100 min at concentrations of 2 μg/ml. Desulfated holothurin A, however, was considerably less effective, taking 190 min at 4 μg/ml to kill the fish. The extract from the sea star, Acanthaster planci, was toxic to killifish in concentrations of 4 μg/ml in 180 min. The extract of the crinoid, Nemaster rubiginosa, was toxic in 150 min at concentrations of 4 μg/ml. The echinoids were either nonichthyotoxic or required very high concentrations of the extract. Extracts of Mellita quinquiesperforata, Clypeaster roseaceus, and the bodies of Diadema antillarum were not lethal at concentrations of 1000 μg/ml, the strongest concentration used. Extracts of the *Diadema* spines were lethal in 200 min at 600 μg/ml and that of *Meoma ventricosa* in 170 min at 400 μg/ml. The extract of the ophiuroid, *Ophiocoma echinata*, caused death in the killifish in 63 min at concentrations of 400 μg/ml.

II. EFFECTS ON SEA URCHIN GAMETES AND DEVELOPMENT

The local sea urchin, *Arbacia punctulata*, is a relatively simple, accessible biological system that has been used extensively in elucidating morphogenetic processes. Sea urchin sperms, eggs, and early developmental stages offer an assay system, which is easy to manipulate (provided the test material is soluble in seawater) and supplies a variety of data in a relatively short period of time.

We have found, for example, that certain substances extracted from marine organisms cause unfertilized eggs to develop parthenogenetically; produce multiple karyokinesis without concomitant cytokinesis; interfere with the dissolving action of the hatching enzyme on the fertilization membrane; affect the synchrony of the ciliary beat; and induce characteristic abnormalities in developmental patterns.

The methods used in this study are the same as those described in previous reports (Ruggieri and Nigrelli, 1960; Ruggieri, 1965).

A. Effects on Sperm Motility

Concentrations of 2 ppm of the extracts from the sea cucumbers *Holothuria parvula*, *Holothuria mexicana*, and *Astichopus multifidus* immobilized sea urchin sperm. Extracts of *Holothuria arenicola* and *Holothuria densipedes* immobilized sperm at concentrations of 10 ppm. Anionic holothurin A immobilized sperm at 5 ppm, whereas 25 ppm of the neutral desulfated holothurin A was needed to arrest sperm motility. A 50 ppm concentration of the extract from the sea star *Acanthaster planci* immediately immobilized sperm as did concentrations of 100 ppm of the crinoid *Nemaster rubiginosa*. The extracts of all the echinoids *Mellita quinquiesperforata*, *Meoma ventricosa*, *Clypeaster rosaceus*, and *Diadema antillarum* and the ophiuroid *Ophiocoma echinata* had no effect on sperm motility at concentrations of 5000 ppm, the highest concentration tested.

B. Effect on Early Developmental Stages (2-Cell Stages)

Depending on the concentration and the length of exposure time, a variety of effects was observed when two-cell stages were subjected to these extracts. For example, relatively strong concentrations for short periods of time or weaker concentrations for longer time intervals caused an arrest in development and cytolytic effects.

The most characteristic abnormality was animalization - a hyperdevelopment of ectodermal structures at the expense of entomesodermal structures. An animalized larva fails to gastrulate and exhibits excessive development of the ciliary tuft, thickening of the apical ectoderm, and absence of archenteron and skeletal spicules. Inhibition of hatching was also a common feature produced by these extracts. Normally the blastula releases a hatching enzyme which completely dissolves the fertilization membrane and the blastula emerges as free-swimming. Varying degrees of interference with the action of the enzyme on the membrane have been observed - from individuals that are able partially to dissolve the membrane and become free-swimming to those which are unable to dissolve the membrane at all. These latter embryos undergo gastrulation and develop further within the membrane without ever emerging as free-swimming.

The extracts of all five sea cucumbers produced animalized larvae at concentrations of 1-10 ppm. These same concentrations also resulted in various manifestations of inhibition of hatching. Anionic holothurin A elicited these effects at concentrations ranging from 0.1 to 5 ppm, whereas desulfated holothurin A was effective at 1 to 10 ppm. These same effects were elicited by extracts of the sea star *Acanthaster planci* at 5 to 50 ppm, and 10 to 100 ppm of the crinoid extract. Extracts of the echinoids and the ophiuroid were effective at concentrations of 500 to 1000 ppm.

Therefore, brief exposure of sea urchin eggs to relatively weak concentrations of extracts from holothurians, asteroids, and crinoid induced animalization. Much stronger concentrations, however, were necessary to elicit this effect with extracts from echinoids and one ophiuroid.

We have shown previously that several other extracts from the Bahamian sea cucumber *Actinopyga agassizi* (Ruggieri and Nigrelli, 1960) and from a number of sea stars (Ruggieri, 1965), such as the above extracts, induced animalization in the developing sea urchin. This abnormality does not appear to be due to the high surface activity exhibited by these extracts. Other surface active agents, for example, quillaja saponin, ouabain, and detergents, have not elicited this effect. The high surface activity of these extracts, however, may play an important role by engendering accumulation at vital surfaces and interfaces.

The possible factors responsible for the induction of animalization have been extensively reviewed by Ranzi (1951, 1957) and Lallier (1958, 1964, 1966). Ranzi's studies on the action of various animalizing agents on the viscosity of proteins led him to conclude that animalizing agents act by denaturing proteins within the developing embryo. Those substances, which prevent denaturation, result in vegetalization or a hyperdevelopment of entomesodermal structure at the expense of ectodermal structures, i.e., the production of exogastrulae.

Lallier (1964, 1966) suggests that animalizing and vegetalizing agents modify the configuration of proteins while they are being synthesized. These modified proteins would be manifested in the hyperdevelopment of ectoderm (animalization) or in the excessive development of the entomesoderm (vegetalization). It would seem, then, that the echinoderm extracts have an effect on normal protein expression in the developing sea urchin, whether by denaturation of proteins already constituted or by modifying proteins during synthesis is not known.

The effect of these extracts in preventing the hatching enzyme from dissolving the fertilization membrane does appear to be due to their surface active properties. This inability to dissolve the membrane occurred in varying degrees. Some blastulae never emerged as free-swimming; others emerged and left incompletely dissolved membranes on the bottom of the dish as evidence of some difficulty in hatching. There is obviously some interference with the reaction of the hatching enzyme on the membrane. It does not appear likely, however, that the enzyme is being inactivated. It would seem more probable that a film forms on the membrane and thus interferes with the activity of the enzyme on its substrate.

III. EFFECTS ON TISSUE CULTURE CELLS

Only the following extracts have been tested thus far on KB oral carcinoma cells and Gray Seal Kidney Cells (GSK): Holothurin A, *Holothuria mexicana*, *Astichopus multifidus*, *Acanthaster planci*, and *Nemaster rubiginosa*.

The extract from the sea star Acanthaster planci in concentrations of 50-100 μg/ml caused a variety of cytological changes in these cell lines. These changes have been characterized as cytoplasmic granulation, cytoplasmic vacuolation, precipitation crystals, cytoplasmic bridges, nuclear budding, nucleolar budding, nuclear fragmentation, increased cytoplasmic RNA, and finally cell lysis. Extracts of Astichopus multifidus caused changes similar to those produced by Holothurin A at concentrations of 100-500 μg/ml. These included: granulation, cytoplasmic bridges, cytoplasmic vacuolation, increased cytoplasmic RNA, and cell lysis. Extracts of the sea cucumber Holothuria mexicana produced cytoplasmic vacuolation and increased cytoplasmic RNA at concentrations of 100 μg/ml and lysis at 750 μg/ml. Extracts of the crinoid Nemaster rubiginosa caused cytoplasmic bridges, cytoplasmic vacuolization, and granulation at 50 μg/ml and lysis at 100 μg/ml (Cecil et al., 1968; Cecil and Nigrelli, 1971).

IV. ANTIFUNGAL EFFECTS

Several of the extracts were tested against Candida albicans and Saccharomyces cerevisiae. Although none of the test substances showed high activity against these fungi in vitro, their relative activities are noteworthy.

TABLE I

Minimum Inhibitory Concentration of Echinoderm Extracts

	Inhibitory effect (μg/ml)	
Echinoderm extract	Candida albicans	Saccharomyces cerevisiae
Holothurin A	10,000	1,000
Desulfated holothurin A	1,000	100
Holothuria mexicana	5,000	100
Holothuria parvula	1,000	100
Astichopus multifidus	100	50
Acanthaster planci	10,000	100
Mellita quinquiesperforata	No inhibition	No inhibition
Ophiocoma echinata	No inhibition	No inhibition
Nemaster rubiginosa	100,000	50,000

[a]Sabourand agar, 28°C, 66 hr.

Table I shows that the extract from Astichopus multifidus was the most effective against these fungi, and that the extracts from the sea cucumbers were more effective than those from the sea star and crinoid. The extracts from the echinoid and ophiuroid showed no activity against these fungi at concentrations of 100 mg/ml, the strongest concentration used.

Only in tests for antifungal activity has neutral desulfated holothurin A been shown to be more active than anionic holothurin A. Shimada (1969) described an antifungal steroid glycoside, holotoxin, which he isolated from the sea cucumber Stichopus japonicus (Selenka). The infrared spectrum of holotoxin is closely similar to holothurin but holotoxin lacks sulfate. The lack of the sulfate moiety, therefore, is probably responsible for the observed tenfold greater antifungal activity of desulfated holothurin A over anionic holothurin A.

These studies, together with previous studies in our laboratories (Rio et al., 1965; Ruggieri and Nigrelli, 1960; Ruggieri, 1965) and those by Yasumoto et al. (1966), indicate striking differences in the activity of extracts from the five classes of the phylum Echinodermata. Yasumoto et al. (1966) investigated the distribution of saponins in five asteroids, three echinoids, and one each of ophiuroid and crinoid. They found that the extracts of echinoids, ophiuroid, and crinoid exhibited no hemolytic activity, while those of the asteroids showed high activity. Our present studies dealing with the effects of extracts from five species of holothurians, four species of echinoids, and one each of asteroid, ophiuroid, and crinoid on a variety of test organisms show the holothurians, asteroids, and crinoid to be highly active while the echinoids and ophiuroid were inactive.

V. PHYLOGENETIC RELATIONSHIPS OF ECHINODERMS

A number of schemes dealing with the phylogenetic relationships between the five classes of echinoderms have been proposed. Fell (1945, 1948) through his studies on fossils and comparative morphology sees a close relationship between the crinoids, asteroids, and ophiuroids, on the one side, and the holothurians and echinoids, on the other. However, an analysis of the types of larvae encountered in four of these five classes places the echinoids in close relationship to the ophiuroids and the asteroids to the holothurians. Fell (1948), however, in a very detailed treatment of echinoderm larval forms highlights their great plasticity and concludes that phylogenetic deductions should not be based on larval forms alone. Further, he finds it impossible to accept any conclusion which implies that ophiuroids and echinoids are more closely related to each other than to the other classes and that holothurians and asteroids are similarly connected (Fell, 1948).

It now appears, however, that recent studies on the biochemistry of echinoderms may lend support to the relationships seen in early larval development. Bergmann (1962) had indicated and Gupta and Scheuer (1968) have confirmed, that studies on sterols show that the asteroids and holothurians contain Δ^7 sterols, while the echinoids, ophiuroids, and crinoids have Δ^5 sterols. Studies on the distribution of quinone pigments in echinoderms (Singh et al., 1967) also suggest close proximity of ophiuroids and echinoids, on one side, as opposed to holothurians and asteroids, on the other. Gupta and Scheuer (1968) conclude that a close phylogenetic relationship exists between the holothurians and asteroids, on the one hand, and another between the ophiuroids, echinoids, and crinoids, on the other. The studies of Yasumoto et al. (1966), showing that saponins occur only in holothuroids and asteroids, lend support to this contention.

Bolker (1967) places the crinoids by themselves because they contain batyl alcohol and arginine phosphate in common with asteroids and holothurians, but possess Δ^5 sterols as do the ophiuroids and echinoids. Hyman (1955) in her monumental work on the Echinodermata maintains that the holothurians and asteroids are closer to the crinoids than are the ophiuroids and echinoids.

Our studies on the activities of the extracts from members of the five classes of echinoderms on a variety of test organisms show the holothurians and asteroids to be similarly highly active, while the echinoids and ophiuroid are not; and the activity of the crinoid places it closer to the holothurians and asteroids than to the echinoids and ophiuroid.

Most of our information on the biochemistry of echinoderms has come from studies on holothurians and asteroids. Only recently have such studies been expanded to include the ophiuroids, echinoids, and crinoids. Further studies on these other classes of echinoderms, especially additional species of crinoids, should shed more light on the relationships of these five classes to one another, and should reveal other substances with interesting physiological activities.

ACKNOWLEDGMENTS

We express our thanks to Dr. Jack T. Cecil and Dr. Martin F. Stempien, Jr., of the Osborn Laboratories of Marine Sciences. We are also indebted to Dr. Jack D. Chanley, Mt. Sinai Medical and Graduate Schools of the City University of New York, for supplying quantities of Holothurin A and desulfated Holothurin A. These studies were supported in part by Sea Grant Award No. 1-35263 and The Scaife Family Charitable Trusts.

REFERENCES

Bergmann, W., in "Comparative Biochemistry" (M. Florkin and H.S. Mason, eds.), Vol. 3, pp. 144-152. Academic Press, New York, 1962

Bolker, H.I., Nature 213, 904 (1967).

Cecil, J.T., and Nigrelli, R.F., Soc. Inv. Path., Newsletter, 3, 20 (1971).

Cecil, J.T., Ruggieri, G.D., and Nigrelli, R.F., Soc. Inv. Path., Newsletter, 1, 3 (1968).

Chanley, J.D., and Rossi, C., Tetrahedron 25, 1897 (1969a).

Chanley, J.D., and Rossi, C., Tetrahedron 25, 1911 (1969b).

Chanley, J.D., Mezzetti, T., and Sobotka, H., Tetrahedron 22, 1857 (1966).

Fell, H.B., Trans. Proc. N.Z. Inst. 75, 73 (1945).

Fell, H.B., Biol. Rev. Cambridge Phil. Soc. 23, 81 (1948).

Friess, S.L., Durant, R.C., and Chanley, J.D., Toxicon 6, 81 (1968).

Friess, S.L., Durant, R.C., Chanley, J.D., and Fash, F.J., Biochem. Pharm. 16, 1617 (1967).

Friess, S.L., Durant, R.C., Chanley, J.D., and Mezzetti, T., Biochem. Pharm. 14, 1237 (1965).

Friess, S.L., Chanley, J.D., Hudak, W.V., and Weems, H.B., Toxicon 8, 211 (1970).

Goldsmith, E.D., Osburg, H.E., and Nigrelli, R.F., Anat. Rec. 130, 411 (1958).

Gupta, K.C., and Scheuer, P.J., Tetrahedron 24, 5831 (1968).

Hyman, L.H., "The Invertebrates: Echinodermata," Vol. 4. Mc Graw-Hill Book Co., New York, 1955.

Lallier, R., Experientia 14, 309 (1958).

Lallier, R., in "Advances in Morphogenesis" (Abercrombie, M. and J. Brachet, eds.), Vol.3, Chapt.4, pp.147-196. Academic Press, New York, 1964.

Lallier, R., *Année Biol.* 5, 313 (1968).
Lasley, B.J., and Nigrelli, R.F., *Toxicon* 8, 301 (1970).
Lasley, B.J., and Nigrelli, R.F., *Zoologica* 56, 1 (1971).
Nigrelli, R.F., *Zoologica* 37, 89 (1952).
Nigrelli, R.F., and Zahl, P.A., *Proc. Soc. Exp. Biol. and Med.* 81, 379 (1952).
Nigrelli, R.F., and Jakowska, S., *Ann. N.Y. Acad. Sci.* 90, 884 (1960).
Quaglio, N.D., Nolan, S.F., Veltri, A.M., Murray, P.M., Jakowska, S., and Nigrelli, R.F., *Anat. Rec.* 128, 604 (1957).
Ranzi, S., *Experientia* 7, 169 (1951).
Ranzi, S., Publ. No. 48, Amer. Assoc. Adv. Sci., Washington, D.C., 1957.
Ricciutti, M.A., and Damato, A.N., *Circulation* 44, Supplement No.2, II-217 (1971).
Rio, G.J., Stempien, M.F. Jr., Nigrelli, R.F., and Ruggieri, G.D., *Toxicon* 3, 147 (1965).
Ruggieri, G.D., *Toxicon* 3, 157 (1965).
Ruggieri, G.D., and Nigrelli, R.F., *Zoologica* 45, 16 (1960).
Shimada, S., *Science* 163, 1462 (1969).
Singh, H., Moore, R.E., and Scheuer, P.J., *Experientia* 23, 624 (1967).
Styles, T.J., *J. Protozool.* 17, 196 (1970).
Styles, T.J., and Hoshino, S., Personal communication (1971).
Sullivan, T.D., and Nigrelli, R.F., *Proc. Amer. Assoc. Cancer Res.* 2, 151 (1956).
Sullivan, T.D., Ladue, K.T., and Nigrelli, R.F., *Zoologica* 40, 49 (1955).
Yamanouchi, T., *Publ. Seto Mar. Biol. Lab.* 4, 184 (1955).
Yasumoto, T., Tanaka, M., and Hashimoto, Y., *Bull. Jap. Soc. Scient. Fish.* 32, 673 (1966).

Chapter 11

BIOLOGICAL ACTIVITY EXHIBITED BY SEAWEED EXTRACTS

T. L. Senn

Department of Horticulture
Clemson University
Clemson, South Carolina

The variety of plants in the sea covers a great spectrum from microscopic species to the giant kelp. Research involving plants harvested from the sea is worldwide. However, the majority of the research is directed toward nonagricultural applications.

Liquid seaweed products were introduced commercially as early as 1950. The most commonly used is the brown seaweed Ascophyllum nodosum, Order Fucaceae. A British patent for a liquid seaweed extract was made as early as 1912. The nutritional value of these products does not seem to be related to their nitrogen, phosphorous, and potassium content.

Their chelating activity was reported in 1959. Seaweed extracts are mostly applied as foliar sprays, and this was contrary to all concepts of plant nutrition when they were introduced in 1950. Research in the 1950s-1960s proved that foliar feeding was effective. Foliar feeding became orthodox practice in the 1960s. This notable change helped the sale of liquid seaweed products throughout the world.

Research with various extracts from Ascophyllum nodosum was initiated at Clemson in 1959. Greenhouse and field studies with horticultural plants sprayed with seaweed extracts exhibited growth patterns that seemed to be induced by the presence of growth regulators in the seaweed extracts used.

At present, growth-promoting substances in seaweeds are well known. It has been shown that rice seeds germinated more readily when soaked in extracts of a blue-green alga. Aqueous extracts have proved more effective than alcohol extracts. Two papers in 1965 and 1967 by Schiewer extend the range of seaweeds which contain indole compounds. He stated that alkaline hydrolysis released large amounts of auxins. Various reports have been made on the presence of gibberellins.

Much of the research in the United States related to the agricultural uses of seaweed has been conducted at Clemson University (Senn et al., 1961; Senn and Skelton, 1966, 1968; Skelton and Senn, 1966; Sirois, 1966). It has shown that seed germination of some plant species is increased by treatment with seaweed extracts (Senn and Skelton, 1966, 1968). Treatment of seeds of several species with seaweed extracts resulted in greatly accelerated respiratory activity. As the concentrations of seaweed extract were increased, the rate of respiratory activities also increased. At the highest concentrations studied, where the respiratory activity was highest, seed germination was low. Changes in respiratory activities of plants have also been induced through applications of seaweed meal and extracts (Aitken, 1964; Aitken et al., 1961; Senn and Skelton, 1966). These results suggest the presence of a respiratory stimulant in the seaweed meal and extract. This unknown material may be a hormone commonly found in plants or only in marine vegetation. See Addendum.

Fruit harvested from peach trees which had been sprayed with seaweed extracts had increased shelf life when compared to similar fruit from untreated trees (Childers, 1965; Senn and Skelton, 1968; Sirois, 1966). The growth of decay organisms, which may be present on the surface of the fruit during the growing season and following harvest, appeared to be inhibited by treatment with seaweed extracts. Earlier sprays, beginning during bloom, resulted in the best fruit shelf life.

From previous field studies, the changes in plant growth patterns seem to be induced by the presence of growth regulator(s) in the seaweed extracts employed (Aitken and Senn, 1965; Senn et al., 1960).

Several pieces of evidence suggest the presence of a growth hormone of some type in the seaweed extracts. More evidence will be required to definitely say which one(s) is involved. Explanation of the observed responses will depend on the type or class of hormone determined.

The occurrence of auxins and gibberellins in fresh algae have been studied (Mowat, 1964a,b). However, no detailed quantitative measurements were made and no work has been reported on seaweed extracts. Whether these regulators survive the processing method is of interest because the use of fresh seaweed is quite limited, but the use and potential uses of seaweed extracts, be it in the meal form or liquid form, are almost limitless.

The existence of unknown growth regulators in seaweed extracts is another possibility. These unknown regulators can be an entirely new class of regulators, such as those recently discovered from rape pollen (Miller, 1963), or be modified compounds of known regulators. The isolation and identification of either would be of particular interest to plant science and industry.

Blunden and Woods (1969) studied the effects of various carbohydrates in seaweed fertilizers on plant growth. Various effects were demonstrated that suggest that certain sugars present in the extract may serve as additional energy sources for plant growth.

A complete analysis of the seaweed extract should be conducted. In addition to the analyses of all organic matters, carbohydrates, proteins, lipids, and pigments, the chemical identification and quantitation of plant growth regulators shall be the main part of the composition analyses.

Methods of isolation of plant growth regulators are important. Depending on the compounds of interest, different procedures have to be employed. For auxins, solvent extraction has been well documented (Anthony and Street, 1970; Bentley, 1962). Diethyl ether alone or the combination of diethyl ether and absolute methanol seem to be the most commonly used solvents. The extraction needs to be done at low temperature and acidic pH. Preliminary work on the identification of auxins in seaweed extracts has been tried. The extracted substances show several spots consistently on paper chromatography after spraying with Ehrlich reagent (Bentley, 1962). Identification of these compounds has not been achieved.

There exist in nature some bound auxins in addition to free auxins. The bound auxins need to be hydrolized either with enzymes or acid or alkaline reagents, to release the auxins into solution before the ether extraction (Anthony and Street, 1970). Because of the complex nature of seaweed cells, the existence of bound auxins cannot be overlooked.

Methods for the extraction of other growth regulators need to be similarly studied. In the chemical analyses of plant growth hormones, the auxins can be studied by gas chromatographic separation after methylation with diazomethane (Osborne and McCalla, 1961) or by using paper and thin-layer chromatographies coupled with color reactions (Murashige and Skoog, 1962; Phinney, 1957). Gibberellins can be determined by gas chromatography or the combination of gas chromatographic-mass spectrometry techniques after either methylation or trimethylsilylation of the crude extracts (Jones and Varner, 1967). Chemical methods for other growth regulators are not as well established.

I. BIOLOGICAL ACTIVITY

One of the overall objectives of the research project is to determine the type or class of biological activity exhibited by seaweed extracts. Initial emphasis will involve bioassays characteristic for auxins, gibberellins, cytokinins, abscissic acid, and ethylene. Specific objectives are (1) to determine the growth regulating activity of the seaweed extract by using bioassays specific for auxins, gibberellins, cytokinins, abscissic acid, and ethylene; (2) to evaluate the responses observed in standardized growth regulator bioassays in terms of new or unknown compounds if these responses cannot be attributed to a known type of growth regulator.

II. METHODS OF PROCEDURE: BIOLOGICAL ACTIVITY

Initial studies will involve standard or classic bioassays specific for auxins, gibberellins, cytokinins, abscissic acid, and ethylene. Each experimental design will include a range of concentrations of the growth regulator for which the bioassay is specific. The seaweed extract will be tested undiluted (full strength) and at several dilutions (1:1, 1:5, 1:10, and 1:50). Three replications will be used containing three treatments, each to ensure statistical significance. Environmental conditions of each bioassay will be rigidly controlled.

The specific bioassays that will be used are outlined below:

1. Auxins. Straight growth of 4 mm segments of *Avena* coleoptile and mesocotyl will be measured and expressed as either absolute elongation or as a percentage of initial length (Sirois, 1966; Thimann, 1969). Indole-3-acetic acid will be used as a standard for comparison.

2. Gibberellins. Stem dwarfism [dwarf corn (Phinney, 1957) and peas (Hayaski and Rappaport, 1966) and induction of a-amylase in barley endosperm (Jones and Varner, 1967)] will be used. Gibberellic acid (Ga_3) will be included as a standard.

Variations in growth rate will be measured in dwarf mutants. Starch digestion will be used for determination of a-amylase activity.

3. Cytokinins. Induction of cell division in tobacco pith tissue (Murashige and Skoog, 1962) and in soybean callus tissue (Miller, 1963), and chlorophyll preservation tests (Osborne and McCalla, 1961) will be included for cytokininlike activity determination. Kinetin will be used as a standard for comparison.

4. Abscissic acid. Abscission-accelerating or -retarding properties will be studied using explants excised from 14-day-old cotton seedlings. (Bornman et al., 1967). Abscissic acid will be included as a standard for comparison.

5. Ethylene. Induction of epinasty in leaves of tomato and pea plants (Pratt and Goeschl, 1969) and ripening of tomato fruits (Burg and Burg, 1965). Commercially available ethylene will be used as a standard.

The extensive research being conducted in various areas of the country has uncovered some enlightening factors concerning seaweed as a plant stimulant. It seems, however, that much is yet to be done before all the factors responsible for improved growth and quality of plants are known.

ADDENDUM

TABLE I

The Effect of Varying Concentrations of Seaweed Extract on the Subsequent Respiratory Activity of Ligustrum lucidum seed (January 2, 1961)[a]

		CO_2 Evolution (mm^3/gm/hr)				
		1-100 Extract	Check	1-25 Extract	1-5 Extract	Pure Extract
	mm^3 CO_2	37.2	42.9	62.4	114.7	164.6
Treatment	/gm/hr	Difference between means				
1-100 Extract	37.2	–	5.7	25.2*	77.5**	127.6**
Check	42.9		–	19.5*	71.8**	121.7**
1-25 Extract	62.4			–	52.3**	102.2**
1/5 Extract	114.7				–	49.9**
Pure extract	164.6					–

[a]* and ** indicate significance at the 5% and 1% levels, respectively.

TABLE II

The Effect of Varying Concentrations of Seaweed Extract on the Subsequent Respiratory Activity of Nandina domesticum seed (January 2, 1961)[a]

		CO_2 Evolution (mm^3/gm/hr)				
		1-100 Extract	Check	1-25 Extract	1-5 Extract	Pure Extract
Treatment	mm^3 CO_2 /gm/hr	22.8	24.3	41.3	218.8	601.4
			Difference between means			
1-100 Extract	22.8	–	5.5	18.5*	196.0**	579.6**
Check	24.3		–	17.0*	194.5**	577.1**
1-25 Extract	41.3			–	177.5**	560.1**
1-5 Extract	218.8				–	382.6**
Pure extract	601.4					–

[a]* and ** indicate significance at the 5% and 1% levels, respectively.

TABLE III

The Effect of Seaweed Extract on the Subsequent Respiratory Activity of Loblolly Pine Seed (September 30, 1960)[a]

		Check	1-100	1-25	1-5
		mm^3 CO_2 evolved (gm/hr)			
Treatment	mm^3 CO_2 evolved (gm/hr)	17.3	32.2	43.1	61.7
		Difference between means			
Check	17.3	–	14.9*	25.8**	44.4**
1-100	32.2		–	10.9	29.5**
1-25	43.1			–	18.6*
1-5	61.7				–

[a]* and ** indicate significance at the 5% and 1% levels, respectively.

TABLE IV

The Effect of Seaweed Extract on the Subsequent Respiratory Activity of Loblolly Pine Seed (September 30, 1960)[a]

Treatment	Cu. mm. CO_2 evolved (gm/hr)	Check	1-100	1-25	1-5
		mm^3 CO_2 evolved (gm/hr)			
		103.5	191.1	258.5	369.9
		Difference between means			
Check	103.5	–	89.6*	155.0**	266.4**
1-100	193.1		–	65.4	176.8**
1-25	258.5			–	111.4*
1-5	369.9				–

[a]* and ** indicate significance at the 5% and 1% levels, respectively.

TABLE V

The Effect of Seaweed Meal Applications on the Subsequent Respiratory Activity of Aromatic Tobacco Leaves at 30^oC (June 20, 1959)[a]

Application	Means mm^3 CO_2 (gm/hr)	500#A	400#A	300#A	200#A	100#A	Check
		CO_2 Evolution (mm^3 CO_2/gm/hr)					
		275	279	288	323	331	378
		Difference between means					
500#Acre	275	–	4	13	48	56	103*
400#Acre	279	–	–	9	44	52	99*
300#Acre	288	–	–	–	35	43	90*
200#Acre	323	–	–	–	–	9	55
100#Acre	331	–	–	–	–	–	47
Check	378	–	–	–	–	–	–

[a]* and ** indicate significance at the 5% and 1% levels, respectively.

TABLE VI

The Effect of Seaweed Meal Applications on the Subsequent Respiratory Activity of Aromatic Tobacco Leaves at 30°C (July 17, 1959)[a]

Application	Means mm^3 CO_2 (gm/hr)	500#A	400#A	300#A	200#A	100#A	Check
		CO_2 Evolution (mm^3 CO_2/gm/hr)					
		239	252	274	287	316	363
		Differences between means					
500#Acre	239	–	13	35	48	77*	124**
400#Acre	252	–	–	22	35	64	211**
300#Acre	274	–	–	–	13	42	89*
200#Acre	287	–	–	–	–	29	76*
100#Acre	316	–	–	–	–	–	47
Check	368	–	–	–	–	–	–

[a]* and ** indicate significance at the 5% and 1% levels, respectively.

TABLE VII

The Effect of Seaweed Meal Applications on the Subsequent Respiratory Activity of Aromatic Tobacco Leaves at 30°C (August 26, 1959)[a]

Application	Means mm^3 CO_2 (gm/hr)	500#A	400#A	200#A	100#A	300#A	Check
		CO_2 Evolution (mm CO_2/gm/hr)					
		Differences between means					
		283	297	348	370	378	398
500#Acre	283	–	14	65	87*	95*	115**
400#Acre	297	–	–	51	73*	81*	101*
200#Acre	348	–	–	–	22	30	50
100#Acre	370	–	–	–	–	8	28
300#Acre	398	–	–	–	–	–	20
Check	398	–	–	–	–	–	–

[a]* and ** indicate significance at the 5% and 1% levels, respectively.

TABLE VIII

The Effect of Soil Applications of Seaweed Meal on the Subsequent Respiratory Activity of Geranium Leaves. Dry Weight Basis (February 26, 1960)[a]

Application	Means mm^3 CO_2 (gm/hr)	Check 287	5% Seaweed 374	20% Seaweed 409	10% Seaweed 440	40% Seaweed 482
		Differences between means				
Check	287	–	87**	122**	153**	195**
5% Seaweed	374		–	35	66	108**
20% Seaweed	409			–	31	73*
10% Seaweed	440				–	42
40% Seaweed	482					–

[a] * and ** indicate significance at the 5% and 1% levels, respectively.

REFERENCES

Aitken, J.B., M.S. thesis. Clemson University, 1964.

Aitken, J.B., and Senn, T.L., Botanica Marina 8, 144 (1965).

Aitken, J.B., Senn, T.L., and Martin, J.A., Clemson College Dept. of Horticulture, Research Series No. 24 (1961).

Anthony, A., and Street, H.E., New Phytol. 69, 47 (1970).

Bentley, J.A., Methods of Biochem. Anal. 9, 75 (1962).

Blunden, G., and Woods, D.L., Proceedings 6th International Seaweed Symposium p.647 (1969).

Bornman, C.H., Spurr, A.R., and Addicott, F.T., Amer. J. Bot. 54, 125 (1967).

Burg, S.P., and Burg, E.A., Science 148, 1190 (1965).

Childers, J. T., M.S. thesis, Clemson University (1965).

Hayashi, F., and Rappaport, L., Plant Physiol. 41, 53 (1966).

Jones, R.L., and Varner, J.E., Planta 72, 155 (1967).

Miller, C.O., Modern Aspects Plant Anal. 6, 194 (1963).

Mowat, J.A., 4th International Seaweed Symposium (1964a).

Mowat, J.A., Botanica Marina 8, 149 (1964).

Murashige, T., and Skoog, F., Plant Physiol. 15, 473 (1962).
Osborne, D.J., and McCalla, D.R., Plant Physiol. 36, 219 (1961).
Phinney, B.O., Proc. Nat. Acad. Sci. 42, 185 (1957).
Pratt, H.K., and Goeschl, H.D., Ann. Rev. Plant Physiol. 20, 541 (1969).
Senn, T.L., Martin, J.A., Crawford, J.H., and Derting, C.W. Clemson College, Dept. of Horticulture, Research Series No.23 (1961).
Senn, T.L., Martin, J.A., Crawford, J.H., and Skelton, B.J. Proc. Assoc. Sou. Ag. Workers, 57th Ann. Conv. p. 182 (1960).
Senn, T.L., and Skelton, B.J., Clemson University, Dept. of Horticulture, Research Series No. 76 (1966).
Senn, T.L., and Skelton, B.J., Proc. of VI Int. Seaweed Symp. Santiago de Compostela, Spain (1968).
Skelton, B.J., and Senn, T.L., Clemson University Dept. of Horticulture, Research Series No.86 (1966).
Sirois, J.C., Plant Physiol. 41, 1308 (1966).
Thimann, K.V., in "Physiology of Plant Growth and Development" (M.B. Wilkins, ed.) pp. 3-37, McGraw-Hill, New York, 1969.

Chapter 12

DISTRIBUTION OF SEA SNAKES IN SOUTHEAST ASIA AND THE FAR EAST AND CHEMISTRY OF VENOMS OF THREE SPECIES

Anthony T. Tu

Department of Biochemistry
Colorado State University
Fort Collins, Colorado

I. SEA SNAKE COLLECTING

In 1967, sea snakes were collected in Japan, Formosa, Hong Kong, Thailand, Malaysia, and the Philippines, and a total of 2600 sea snakes were captured (Tu and Tu, 1970). In 1969, we captured over 7000 additional sea snakes in Thailand, Malaysia, and the Philippines.

A. Thailand

The 1967 collecting period was July 1 to July 10. The number of snakes captured were: 68 *Lapemis hardwickii*, 11 *Enhydrina schistosa*, 146 *Aipysurus eydouxi*, 1 *Pelamis platurus*, *Hydrophis cyanocinctus* (data lacking), *Hydrophis spiralis* (data lacking), *H. klossi* (data lacking), *Kerilia jerdoni* (data lacking), *Microcephalophis gracilis* (data lacking), and 50 unknown sea snakes. The false sea snake, *Acrochordus granulatus* was captured. *Acrochordus granulatus* is a nonpoisonous snake and, therefore, lacks any fangs. It is reported that they are found in rivers and estuaries (Taylor, 1965). However, we found many of *A. granulatus* in the open sea in the Gulf of Thailand, both in 1967 and 1969. *Aipsysurus eydouxi* was the predominant species during the 10-day period. *Lapemis hardwickii* was the second most abundant sea snake in the 1967 collection.

Enhydrina schistosa is commonly found in the estuary in central Thailand and the snake moves upstream with the tidal influx of seawater. In the vicinity of the Bangkok area, E. schistosa is most abundant during the dry season (December to April), when seawater penetration is greatest owing to the low water level of the rivers. In the Bangkok area, seawater can reach to Phra Pra Dang, which is 3 miles south of Bangkok, or about 10 miles from the river mouth. Fishermen there have reported catches of about 1000 sea snakes per night during the dry season. An attempt was made to collect the sea snakes in the same spot in June 1967, but only two small E. schistosa and three A. granulatus were caught in 1 day.

In 1969, the collecting was made about 10 miles off the east coast of the Kra Isthmus with trawling nets attached to trawlers. Species and number of sea snakes captured are summarized in Table I. The data obtained from the 1969 collection

TABLE I

Record of Sea Snake Collecting in Thailand From July 3 to August 13, 1969

Snake	Number
Lapemis hardwickii	4305
Aipysurus eydouxi	146
Hydrophis cyanocinctus	92
Hydrophis ornatus	73
Enhydrina schistosa	73
Kerilia jerdonii siamensis	55
Praescutata viperina	99
Microcephalophis gracilis	16
Thalassophis anomalus	4
Unknown	165
Acrochordus granulatus (nonpoisonous)	186
Total	5306

are statistically more significant than that of 1967, since a longer period of time was spent for the collection, which was from July 3 to August 13, 1969. Since a large number of sea snakes were captured, the distribution of each species can be fairly accurately projected for the summer period. Of the 5306 sea snakes captured, 4305 were *Lapemis hardwickii*, which accounted for 82% of all the snakes. The next common species was *Aipysurus eydouxi*, which accounted for 4.5%. Again large numbers of *Acrochordus granulatus* (186) were captured in the open sea. This is the only nonpoisonous snake captured in the sea. Contrary to the reported record, *A. granulatus* can live in salt water. It may be that the snake possesses a salt gland so that it can survive in a high tonic environment.

Pelamis platurus is very rare in the Gulf of Thailand. Out of a total of nearly 6000 sea snakes captured in 1967 and 1969 in Thailand, only one was *P. platurus*. Contrary to the report of the Royal Thai Navy, no genus of *Laticauda* was found in Thailand. Sea snakes reported to be found in Thailand and by the Royal Thai Navy are *Lapemis hardwickii*, *L. curtus*, *Enhydrina Schistosa*, *Thalassophis viperina*, *Astrotia stokesii*, *Microcephalophis gracilis*, *Kerilia jerdonii*, *Pelamis platurus*, *Laticauda laticaudata affinis*, *L. colubrina*, *L. semifasciata*, *Emydocephalus ijimae*, *Hydrophis spiralis*, *H. cyanocinctus*, *H. ornatus*, and *H. melanocephalus*. The sea snakes in the Gulf of Thailand reported by Taylor (1965) are *Laticauda colubrina*, *L. laticaudata*, *Aipysurus eydouxi*, *Kerilia jerdoni*, *K. jerdoni siamensis*, *Astrotia stokesii*, *Kolpophis annandalei*, *Thalassophis viperina*, *Enhydrina schistosa*, *Pelamis platurus*, *Lapemis hardwickii*, *Hydrophis cyanocinctus*, *H. ornatus*, *H. caerulescens*, *H. torquatus*, *H. torquatus diadema*, *H. klossi*, *H. fasciatus*, *H. brooki*, and *H. mamillaris*.

B. Philippines

In 1967 and 1969, we collected sea snakes inside the caves of Gato Island. Inside the caves, there was a large number of sea snakes on the surface of the water, in rock crevices, in the water, and on the rock. Sea snakes were captured by skin diving inside the caves. There were two species in this area: One was *Laticauda semifasciata* and the other was *Laticauda colubrina*. Eight-hundred *L. semifasciata* were captured in 1967 and 600 in 1969.

Sea snakes present in the Sea of Philippines reported by Taylor (1922) are *Aipysurus eydouxi*, *Laticauda laticaudata*, *Laticauda colubrina*, *L. semifasciata*, *Hydrophis fasciatus*, *H. ornatus*, *H. cyanocinctus*, *Lapemis hardwickii*, and *Pelamis platurus*. Specimens in the National Museum, Manila, observed by the author, were *Hydrophis fasciatus*, *H. spiralis*, *Pelamis* platurus, and Lapemis hardwickii. Sea snakes identified from the specimens collected by Alaban Serum and Vaccine Laboratories were Lapemis hardwickii, Hydrophis cyanocinctus, and Hydrophis fasciatus atriceps.

C. Malaysia

In Malaysia, sea snakes were captured in the Strait of Malacca near Penang Island by setting fish traps, which were immersed in the water for 12 hr. Usually, at least one snake was caught in each trap. The variety of sea snakes captured was very similar to the ones we obtained in Thailand.

Sea snakes present in the coastal water of Sarawak (in Borneo) are Laticauda laticaudatus, L. colubrina, Aipysurus eydouxi, Kerilia jerdoni, Enhydrina schistosa, Hydrophis cyanocinctus, H. spiralis, H. melanosorana, H. caerulescens, H. torquatus, H. brooki, H. fasciatus, Thalassophis anomalus, Lapemis hardwickii, Microcephalophis gracilis, Pelamis platurus, and Praescutata viperina (Haile, 1958).

D. Other Areas

The most common sea snake in the vicinity of Hong Kong is Hydrophis cyanocinctus, which accounts for 70% of all the sea snakes captured by fishermen in Hong Kong (Romer, 1965). Other sea snakes frequently captured are Hydrophis ornatus ornatus, Microcephalophis gracilis, Pelamis platurus, and Praescutata viperina (Romer, 1961).

In the summer of the 1967 collection, Hydrophis cyanocinctus, H. ornatus, Pelamis platurus, Microcephalophis gracilis, Lapemis hardwickii, and one unknown species were captured.

In 1967, we collected 690 Pelamis platurus from the northern coast of Formosa. In southwestern Formosa, we obtained a large number of genus Hydrophis, however, no venom was extracted since there were too many different species and subspecies within this genus, which made proper identification very difficult.

In Amami Island, Japan, we obtained 500 Laticauda semifasciata. Comparison to those of Philippine origin is made later in this chapter.

II. BIOLOGY

A. Lapemis hardwickii

Unlike the genus Hydrophis, Lapemis hardwickii can be identified quite easily. The length is relatively short for the width of the body. The specimens stored at Colorado State University have the length of 75 and 64 cm, respectively. Because of its short length, L. hardwickii gives the appearance of being a stout snake. Five specimens of L. hardwickii are currently stored in my laboratory and scale rows of the two specimens are reported here. There are 26 scales around the neck for the specimen 69-1 and 33 for specimen 69-2. Transverse scale row for specimen 69-1 is 146, and for specimen 69-2 it is 185. Subcaudal scale for one specimen is 41, while the other gives 46. The scale patterns of the head for dorsal and lateral views are described in Figs. 1A and B. Ventral scale pattern is shown in Fig. 1C. The ventral scale is the largest and gives the appearance of a somewhat flattened hexagonal shape. In many instances there is a tubercle at the center of each scale. The tubercle is the largest in the ventral scales. Young L. hardwickii may not have keeled scales or they may have small tubercles which can be seen only with extremely careful observation. Similar to all other sea snakes, L. hardwickii also has a flattened oarlike tail.

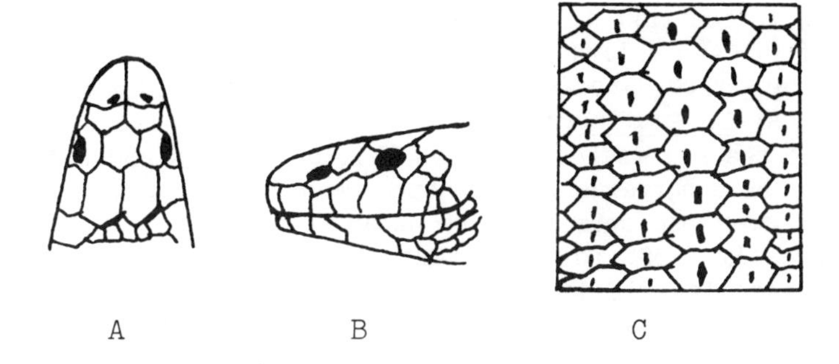

FIG. 1. (A) Head of Lapemis hardwickii from dorsal view. (B) Head of Lapemis hardwickii from lateral view. (C) Appearance of scales on the ventral side. Notice the keeled scales.

B. Laticauda semifasciata

At Amami Island in 1967, 500 specimens were obtained, which the snake handler informed us were captured in Okinawa. More than 90% of the specimens had three prefrontal scales and two internasal scales. About 2 to 3% of the specimens had three internasal scales (see Fig.2B). The other 2 or 3% of the specimens had two prefrontal scales instead of three (Fig. 2C). Venoms were extracted separately from the specimens of different varieties. It is not clear whether or not snakes of slightly different scale patterns have any different chemical properties in their venoms.

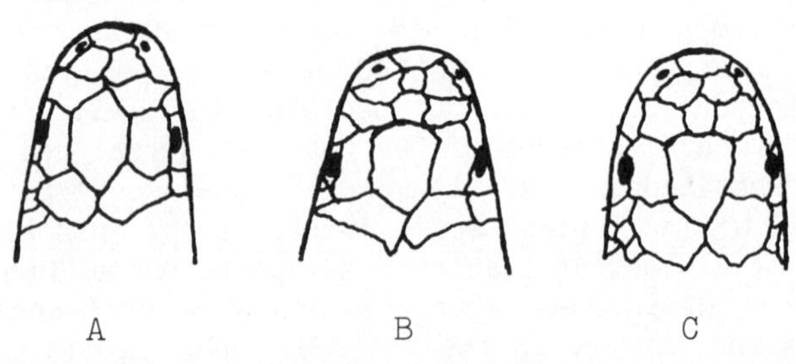

FIG.2. Variation of head scale patter of Laticauda semifasciata captured in Japan in 1967.

The size and length of specimens obtained in Japan were quite small. The average length of the sea snakes was about 3 to 4 ft. In contrast to the specimens from Japan, the snakes captured in Gato Island were very large. Even the smallest L. semifasciata from Gato Island, Philippines was far larger than the largest L. semifasciata from Japan (Fig.3). The average length of the sea snakes from the Philippines was about 7 ft. However, scale patterns for the specimens from Gato Island are identical to those obtained from Japan (Fig.2A).

In early 1920, Taylor investigated the snakes from the Philippines (Taylor, 1922). He described the maximum length of Laticauda semifasciata to be only 150 cm. Therefore, the specimens obtained in Gato Island in 1967 and 1969 may belong to a subspecies or may be an entirely new species. Prior to proper identification by herpetologists, the specimens from Gato Island were designated as Laticauda semifasciata.

There are several chemical differences in the venoms of snakes, L. semifasciata, obtained from the Philippines and from Japan. The chemistry of sea snake venom obtained from the Philippines was studied by Hong (1971), Tu and Hong (1971), and

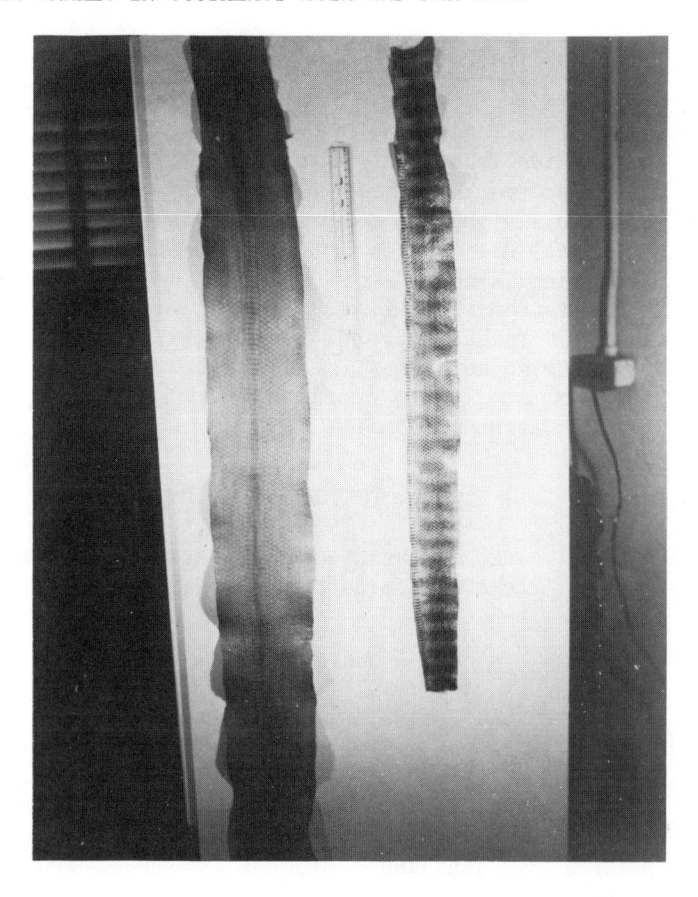

FIG. 3. (A) Skin from Laticauda semifasciata from Gato Island, Philippines captured in 1969. (B) Skin from Laticauda semifasciata from Japan captured in 1969.

Hong and Tu (1970), while that of Japan was studied by Japanese scientists (Arai et al., 1964; Uwatoko et al., 1966a, b; Tamiya and Arai, 1966; Uwatoko-Setoguchi, 1970). These observed chemical differences are summarized here (a) Crystal shapes of toxins a and b from Philippine origin are different from crystal laticatoxin III and crystal laticatoxin IV (Uwatoko et al., 1966b) and erabutoxin a and erabutoxin b from Japan (Tamiya and Arai, 1966). (b) Toxins a and b from the Philippines are stable at 100°C for 30 min, whereas erabutoxin a lost 70% toxicity in 5 min at 100°C and erabutoxin b lost 70% toxicity in 10 min at 100°C. (c) Amino acid compositions of toxins isolated.

from the venoms of the two regions are different. (d) Venom from Japan contains five toxic fractions (Uwatoko et al., 1966a) or four fractions (Tamiya and Arai, 1966) while that of Philippines contains two fractions (Tu and Hong, 1971).

These chemical differences in venoms confirm well the morphological difference of sea snakes in the two regions. The study of chemical differences in the venoms of snakes from different localities is most fascinating. The small chemical differences found from the chemical studies will contribute to the taxonomy of snakes. Miranda et al. (1970) found a small difference in amino acid composition for the cobra venoms from two different sources. They concluded that each venom originated from a different subspecies.

III. TOXICOLOGY

In general, venoms of sea snakes are more toxic than those of land snakes. the LD_{50} in mice by IV is listed in Table II.

TABLE II

Yield and Toxicity of Sea Snake Venoms

Venom	Origin	Yield (mg/snake)	LD_{50} (μg/g)	Year Collected
Aipysurus eyedouxi	Thailand	0.6	>4	1967
Enhydrina schistosa	Malaya	–	0.090	1967
			0.098	1969
	Thailand	8.1	0.14	1967
		14.0	0.21	1969
Hydrophis cyanocinctus	Hong Kong	2.1	–	1967
	Malaya	–	0..35	1967
	Thailand	18.0	–	1969
H. ornatus	Thailand	19.0	2.2	1969
Lapemis hardwickii		5.2	0..71	1967
		2..4	1.40	1969
Laticauda semifasciata	Japan	7.10	0.28	1967
	Philippines	16.0	0.28	1967
		19.0	0.45	1969
Pelamis platurus	Formosa	2.0	0.18	1967
		9.3	0.28	1968

The quantity of venom that can be obtained from sea snakes is much smaller than the amount that can be obtained from land snakes. The yield of venom from sea snakes collected was from 0.6 to 19 mg per snake.

IV. CHEMISTRY

A. Comparison of Venom from Sea Snake and Land Snake

Recent studies have shown that venom from the snakes of the family Hydrophiidae are much simpler in composition than the venom of the land snakes (Toom et al., 1969). Distribution

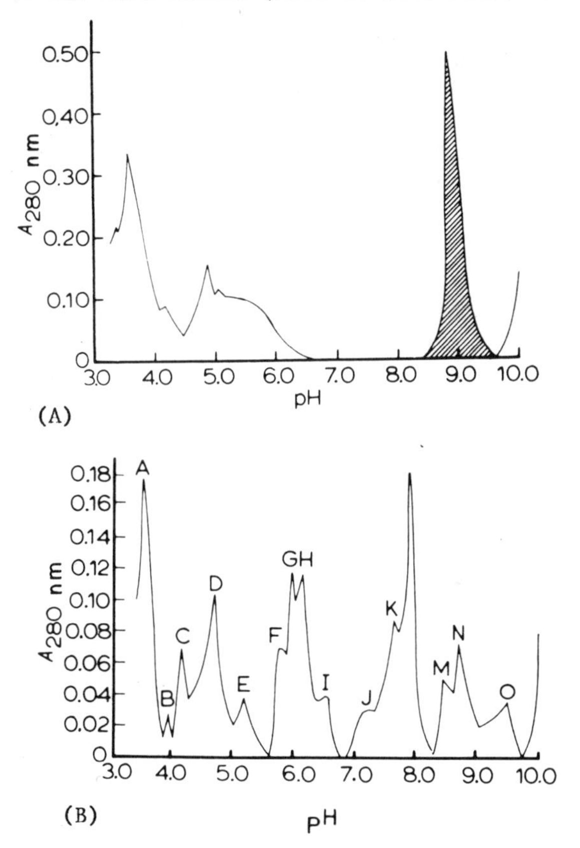

FIG. 4. Comparison of venoms of sea and land snakes. (A) Isoelectric focusing pattern of Enhydrina schistosa venom, obtained from the Strait of Malacca. (B) Isoelectric focusing pattern of Agkistrodon rhodostoma venom, obtained from Malaya.

of proteins in venoms of Enhydrina schistosa and Agkistrodon rhodostoma (Malayan pit viper) is illustrated in Fig.4.

Isoelectric points of sea snake venom toxins are very basic; they are all around or above 9, while that of a land snake, A. rhodostoma, has an isoelectric point of about 7.0.

B. Isolation

Since all snake venoms contain a rather large number of proteins, purification is achieved by more than two-step column chromatography. Toxins were isolated either using the combination of Sephadex G-50 and CM-cellulose chromatography (Tu et al., 1971; Tu and Hong, 1971) or repeating use of CM-cellulose with a different buffer (Tu and Toom, 1971). Venoms used for isolation were Lapemis hardwickii from Thailand, Laticauda semifasciata from the Philippines, and Enhydrina schistosa from Malaysia. Phospholipase A was isolated from venom of L. semifasciata from Amami Island, Japan by a two-step purification utilizing CM-cellulose and DRAE column chromatography (Tu et al., 1970).

C. Criteria of Purity

In each preparation, three or four of the following methods were used to confirm the purity of isolated toxins or the enzyme.

a. Sedimentation pattern in analytical ultracentrifuge
b. Polyacetate electrophoresis at different pH values
c. Isoelectric focusing
d. Rechromatography in column
e. Straight line in the plot of log C against r^2 in sedimentation equilibrium
f. Crystallization

D. Physical and Chemical Properties

Physical and chemical properties of isolated toxins and phospholipase A are summarized in Table III.

The toxins contain a total amino acid residue of either 61 or 62. The phospholipase A consists of 108 amino acid residues.

Molecular weights of toxins and phospholipase A were determined by a combination of s and D, data from sedimentation equilibrium, amino acid composition, or quantitative Sephadex elution method. These results are summarized in Table IV.

TABLE III

Physicochemical Properties of Sea Snake Venom Toxins and Phospholipase A

	Enhydrina[a] schistosa (Malaysia) toxin	Lapemis[b] hardwickii (Thailand) toxin	Laticauda[a] semifasciata (Philippines) toxin		Laticauda[a] semifasciata (Japan) phospholipase A
			a	b	
Isoelectric Point	9.20	9.85	9.15	9.34	6.70
Sedimentation coefficient ($s^{o}_{20,w}$)	1.4	1.13	1.52	1.43	1.93
Diffusion coefficient ($D^{o}_{20,w}$ cm^2/sec)	15.5 x 10^{-7}	13.7 x 10^{-7}	–	–	14.1 x 10^{-7}
Parital specific volume	0.70	0.70	0.71	0.71	0.71
Amino terminal	–	His	Arg	Arg	–
Carboxy terminal	–	Glu	Asp	Asp	–
Amino acid residue	61	62	62	61	108

[a] Venom was collected in 1967
[b] Venom was collected in 1969

End-group analysis indicated that there is a histidine at the amino terminal and aspartic acid or asparagine at the carboxy terminal for the toxin of Lapemis hardwickii venom. For toxins a and b of Laticauda semifasciata venom from the Philippines, the amino terminal is arginine and the C terminal is aspartic acid (asparagine).

Amino acid compositions of purified toxins and phospholipase A are summarized in Table V. All toxins contain 8 moles cysteine, 8 moles glutamic acid, 1 mole each of tryptophan, leucine, and tyrosine. It is remarkable that there is a common number of residues for certain amino acids regardless of geographical origin. Other amino acid compositions are also remarkably similar. Only the toxin from Lapemis hardwickii contains methionine. Amino acid composition of phospholipase A is quite different from those of other toxins.

TABLE IV

Molecular Weight of Sea Snake Venom Toxins and Phospholipase A

Sea snake	Origin	Amino acid composition	S & D	Sedimentation equilibrium	Gel filtration	Reference
Lapemis hardwickii toxin	Gulf of Thailand	6774	6800	6800	–	Tu and Hong (1971)
E. schistosa toxin	Strait of Malacca	6878	7300	–	–	Tu and Toom (1971)
Laticauda semifasciata						
toxin a	Gato Island, Philippines	6840	–	6800	6600	Tu et al. (1971)
toxin b		6677	–	6500	6400	
Laticauda semifasciata						
phospholipase A	Amami Island, Japan	–	11000	–	10700	Tu et al. (1970)
		–	–	–	10400	

E. Chemical Modification

The tryptophan residue was chemically modified using a specific reagent, N-bromosuccinimide, on toxins isolated from the venoms of Laticauda semifasciata, Enhydrina schistosa, and Lapemis hardwickii (Tu and Toom, 1971; Tu et al., 1971; Tu and Hong, 1971). These toxins contain only 1 mole of tryptophan residue. After the modification, the toxicity disappeared completely. Two other reagents, 2-nitrophenylsulfenyl chloride and 2-hydroxyl-5-nitrobenzyl bromide, were used for the modification of tryptophan residue of the toxin isolated from the venom of Lapemis hardwickii (Tu and Hong, 1971). Toxicity of the toxin disappeared again on modification. It is thus concluded that the tryptophan residue is important for toxic action.

In contrast to tryptophan, the modification of the majority of arginine and lysine residues did not alter the toxicities of toxin a and b of Laticauda semifasciata venom. The result of chemical modification of sea snake venom toxins is summarized in Table VI.

TABLE V

Amino Acid Composition of Sea Snake Venom Toxins and Phospholipase A

	Lapemis hardwickii toxin	E. schistosa toxin	Laticauda semifasciata (Philippines) toxin		Laticauda semifasciata (Japan) phospholipase A
			a	b	
Lysine	5	5	4	5	7
Histidine	2	2	1	1	2
Arginine	3	3	3	2	4
Aspartic acid	6	6	5	4	11
Threonine	8	8	6	5	6
Serine	6	6	7	6	7
Glutamic acid	8	8	8	8	9
Proline	3	3	4	4	5
Glycine	4	5	5	6	10
Alanine	1	1	0	0	8
Valine	1	1	2	3	4
Methionine	1	0	0	0	1
Isoleucine	2	2	4	4	3
Leucine	1	1	1	1	5
Tyrosine	1	1	1	1	10
Phenylalanine	0	0	2	2	3
Half-cystine	8	9[a]	8	8	12
Tryptophan	1	1	1	1	1
Total residue	61	62[a]	62	61	108

[a] Nine residues of cysteine and a total residue of 62 were found by a calculation based on the average molar ration to leucine, alanine, and valine. Eight residues for cysteine and 61 total residues were found if the calculation was based on the ratio to leucine alone. Leucine yielded the smallest number of moles in the amino acid analysis.

F. Absence of Enzyme in Purified Toxins

The venom of E. schistosa contains a number of enzymes. Sixteen substrates were used to test various enzyme activities (Tu and Toom, 1971). The venom shows the following enzyme activities: clotting activity, hyaluronidase, alkaline phosphatase, phosphodiesterase, deoxyribonuclease, acetylcholinesterase, and leucine aminopeptidase. However, the venom does not contain such enzyme activities as ribonuclease, acid phosphatase, amino acid esterase (with N-benzoyl-L-arginine ethyl ester, N-benzoyl-L-tyrosine ethyl ester, p-toluenesulfonyl-L-arginine methyl ester, and acetyl-L-tyrosine ethyl ester as substrates), and proteases (with casein and hemoglobin as substrates)) None of the above enzymes are found in purified toxins.

TABLE VI

Summary of Chemical Modifications

Toxin	Amino acid	Reagent	Toxicity	Amino acid residue before modification	After modification	Number of amino acid residue modified
Laticauda semifasciata (Philippines)						
toxin a	Tryptophan	N-bromosuccinimide	–	1	0	1
	Arginine	1,2-Cyclohexanedione	+	3	2	1
	Lysine	O-Methylisourea	+	4	1	3
toxin b	Tryptophan	N-Bromosuccinimide	–	1	0	1
	Arginine	1,2-Cyclohexanedione	+	2	1	1
	Lysine	O-Methylisourea	+	5	1	4
Enhydrina schistosa (Malaysia)	Tryptophan	N-Bromosuccinimide	–	1	0	1
Lapemis hardwickii (Thailand)	Tryptophan	2-Nitrophenylsulfenyl chloride	–	1	0	1
		2-Hydroxy-5-nitrobenzyl bromide	–	1	0	1
		N-Bromosuccinimide	–	1	0	1

[a] + Toxic
[b] – Nontoxic

G. Properties of Phospholipase A

By using ovolecithin of known composition in the 1 and 2 positions, the enzyme is shown to be specific for the 2 position, liberating mainly unsaturated fatty acids (Tu et al., 1970).

Of the substrates tested, only phosphatidylcholine was hydrolyzed. Of the two phosphatidylcholines tested, ovolecithin was hydrolyzed at a much more rapid rate than the synthetic lecithin containing only the saturated fatty acid, palmitic acid. All other substrates tested, namely, phosphatidyl ethanolamine, phosphatidyl-L-serine, phosphatidyl inositide, phosphatidic acid, lysolecithin, sphingomyelin, cerebromide, and cardiolipin were not hydrolyzed. The enzyme was most active at pH 8.0 and at temperatures between 35° and $40^{\circ}C$. The activation energy calculated from Arrhenius plot is 6900 cal/mole.

The enzyme exhibited hemolytic activity which was greatly intensified by the addition of lecithin (Tu et al., 1970). The purified phospholipase A was nontoxic, nonhemorrhagic, and exhibited only slight myolytic activity. It was found that phospholipase A even in presence of ovolecithin had very little effect on the mouse embryo cells in tissue cultures.

V. IMMUNOLOGY

Neutralization capacity of commercial antivenin (Commonwealth Serum Laboratories, Melbourne, Australia) in vitro was tested against homologous and heterologous venoms (Tu and Ganthavorn, 1969). The antivenin was not only effective for homologous venom, but it also effectively neutralized three heterologous venoms tested. One milliliter of serum neutralized 176 times the LD_{50} value for its own venom, 160 LD_{50} value for *Pelamis platurus* venom from Formosa, and 120 LD_{50} value for the venoms of *Hydrophis cyanocinctus* from Malaya and *Lapemis hardwickii* from Thailand.

Immunodiffusion and immunoelectrophoretic studies indicate that different sea snake venoms contain some common antigens (Figs.5-8).

VI. DISCUSSION

Low molecular weights and high contents of disulfide bonds (8) of toxins for all sea snake venoms suggest that these toxin molecules have a rather compact structure, thus probably ac-

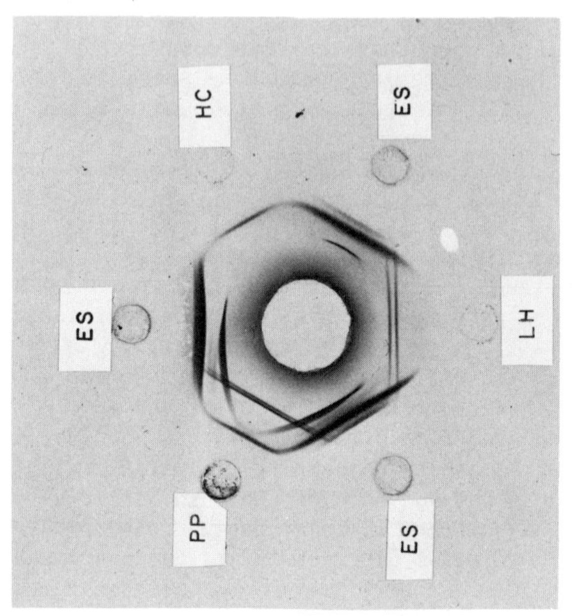

FIG. 6

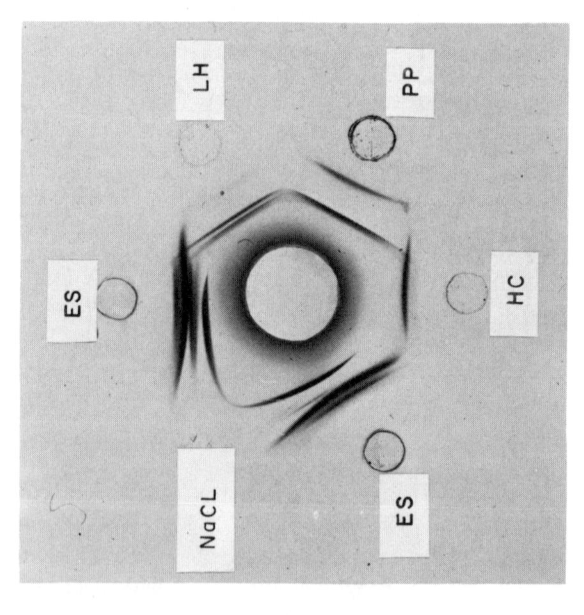

FIG. 5

Figs. 5, 6 and 7. Immunodiffusion pattern of sea snake (ES, LH, PP and HC) venom against antibody of ES. ES: Enhydrina schistosa; LH: Lapemis hardwickii; PP: Pelamis platurus; and HC: Hydrophis cyanocinctus.

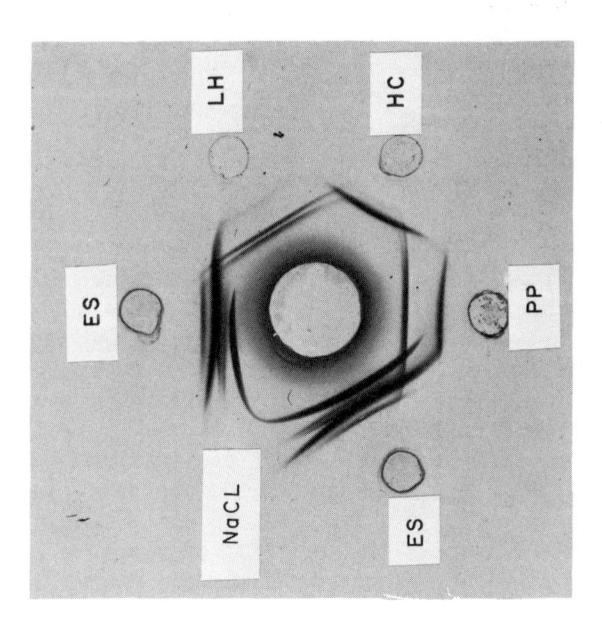

FIG. 7

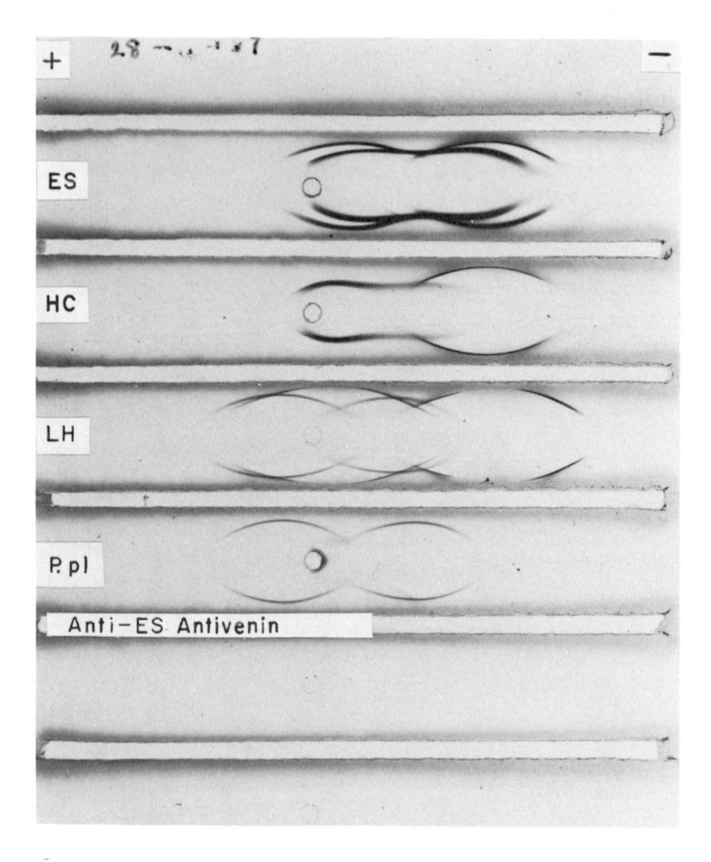

FIG. 8. Immunoelectrophoretic patterns of anti-ES venom against different sea snake venoms.

counting for the high stability of toxins at high temperature and over a wide range of pH. Our findings seem to confirm high thermostability and high toxic activity against a wide range of pH for the venoms of Laticauda semifasciata, L. laticaudata, and Hydrophis cyanocinctus from Japan (Homma et al., 1964). However, low heat stability for erabutoxins a and b were reported (Tamiya and Arai, 1966).

It is most interesting that chemical properties of toxins from sea snakes regardless of geographical origin are strikingly similar. They either contain 61 or 62 amino acid residues, only 1 tryptophan, leucine, or tyrosine residue, and 4 disulfide bridges. These data are summarized and compared in Table VII. It is more fascinating that sea snake venom toxins

TABLE VII

Purified toxins from the Venom of Sea Snakes
(Family: Hydrophiidae)

Name	Origin	Nomenclature of toxins	Total amino acid residues	Reference
Lapemis hardwickii	Thailand	Toxin	61	Tu and Hong (1971)
Enhydrina schistosa	Malaysia	Toxin	62	Tu and Toom (1971)
Laticauda semifasciata	Philippines	Toxin a	62	Tu et al. (1971)
		Toxin b	61	
Laticauda laticaudata	Japan	Laticotoxin a	62	Sato et al. (1969)
Laticauda colubrina	Japan	Laticotoxin a	62	Sato et al. (1969)
Laticauda semifasciata	Japan	Erabutoxin a	62	Tamiya and Arai (1966)
		Eratutoxin b	62	Tamiya and Arai (1966)
		Erabutoxin c	–	Uwatoko (1970)
		Laticatoxin III	–	Uwatoko et al. (1966b)
		Laticatoxin IV	–	Uwatoko et al. (1966b)

are not only similar to each other, but also are similar to the cobra toxins (Table VIII). In the case of cobra toxins, the number of amino acid residues is not only 61 and 62, but also 71.

Therefore, all the toxins from sea snake venoms and cobra venom with 61 and 62 amino acid residues should be considered as a homologous toxin and be called by a unified nomenclature in the future.

The carboxy terminals of toxins for Lapemis hardwickii and Laticauda semifasciata (Philippines) venoms are either aspartic acid or asparagine (Tu and Hong, 1971; Tu et al., 1971). The amino terminal for Lapemis hardwickii venom is histidine whereas that of Laticauda semifasciata (Philippines) is arginine. The carboxyl terminals for the erabutoxins a and b are identified to be asparagine, and the amino terminals are arginine (Endo et al., 1971).

The importance of the tryptophan residue for the lethal action of purified toxins from the venoms of Enhydrina schistosa from the Strait of Malacca, Lapemis hardwickii from the Gulf

TABLE VIII

Purified Toxins from Venoms of Elapidae

Name	Origin	Nomenclature of toxins	Total amino acid residues	Reference
Naja naja	India	Cobramine	52	Larsen and Wolff (1968)
		Toxin 3	71	Karlsson et al. (1971)
		Toxin 4	71	
		Toxin A	61	Nakai et al. (1970)
Naja naja atra	Formosa	Cobrotoxin	62	Yang (1965)
		Cardiotoxin	60	Narita and Lee (1970)
		Fraction IX, X, XII	–	Lo and Chang (1967)
Naja naja siamensis	Thailand	Toxin 3	71	Karlsson et al. (1971)
Naja nigricollis	Africa	Toxin α	61	Karlsson et al. (1966)
Naja haje	N. Africa	Toxin α	61	Botes and Strydom (1969)
		Toxin I, II, III	61	Miranda et al. (1970)
Naja nivea	Africa	Toxin Δ	61	Botes (1970)
		Toxin β	61	
		Toxin α	71	
Hemachatus hemachatus	Africa	Peak 3	61	Porath (1966)
		Peak 5	61	
		Peak 12	60	
		Toxin 2	61	Botes (1970)
		Toxin 4	61	
		DLF	57	Aloof-Hirsch et al. (1968)
Bungarus multicinctus	Formosa	a-Bungarotoxin	74	Mebs et al. (1971)

of Thailand, and Laticauda semifasciata from Gato Island, Philippines has been well demonstrated (Tu and Toom, 1971; Tu et al., 1971; Tu and Hong, 1971). The importance of this particular amino acid residue for other venoms has also been reported (Chang and Hayashi, 1969; Seto et al., 1970). Certainly other amino acid residues must also be involved in the lethal action of the toxins. Since all toxins contain a common amino acid residue of one tyrosine, this amino acid residue may be necessary for toxic action. For the case of Laticauda semifasciata from the Philippines, the modification of the majority

TABLE IX

Fatty Acid Composition of Phospholipase A Hydrolysis Products of Egg Lecithin

	Specific fatty acid (mole%) [a]							
	14:0	14:1	16:0	16:1	18:0	18:1	18:2	20:4
Lecithin standard								
Total fatty acids	Tr[b]	Tr	36.2	1.4	13.8	34.0	12.9	1.6
Lecithin blank[c]								
Hydrolyzed fatty acids	0	0	0	0	0	0	0	0
Nonhydrolyzed fatty acids	Tr	Tr	34.2	1.4	14.6	32.8	14.1	2.7
Lecithin + venom								
Hydrolyzed fatty acids[d]	Tr	Tr	10.7	1.6	4.1	58.8	22.4	2.4
Nonhydrolyzed fatty acids[e]	Tr	Tr	52.9	2.8	21.7	18.8	3.6	0.0
% Hydrolysis			9.1	17.6	8.2	60.0	75.0	100.0
Lecithin + phospholipase A								
Hydrolyzed fatty acids	Tr	Tr	7.8	1.5	2.6	60.2	25.8	2.0
Nonhydrolyzed fatty acids	Tr	Tr	59.3	2.8	23.5	12.6	1.6	0.0
% Hydrolysis			6.2	23.0	6.6	70.0	88.2	100.0

[a] Notation: number of carbon atoms followed by number of double b
[b] TR, trace.
[c] All reactants except phospholipase A, incubated at 23°C, pH 8.0 30 min.
[d] Fatty acids hydrolyzed from lecithin by action of phospholipase Numbers represent the average of three determinations.
[e] Fatty acids remaining attached to lecithin and lysolecithin aft drolysis with phospholipase A; numbers represent the average of determinations.

of lysine residues did not alter the toxicities of toxins a and b. We may conclude that three of four lysine residues in toxin a and four of five in toxin b are not essential for toxic action. Whether the last one-lysine residue, which is resistant to modification, is important or not has not been identified yet. When arginine was modified in toxin a and b, only one of each was modified. We can conclude that at least one of the arginine residues in toxin a and b is not essential for toxic action.

Involvement of the tryptophan residue in biological activity is not unique to snake venom toxins. Chao and Einstein (1971) reported that the modification of the tryptophan residue eliminated the encephalitogenic activity of the protein. It was also demonstrated that the tryptophan was important in nerve growth factor isolated from *Naja naja* venom (Angeletti, 1970).

While the tryptophan residue is essential for toxic action, it is not essential for antigenic action. In the immunodiffusion study, the modified and nonmodified toxins found a single precipitin band. This strongly suggests that there is no change in antigenicity after modification.

The toxin may act in a nonenzymatic fashion. When enzyme activities such as clotting activity, alkaline and acid phosphatase, acetylcholinesterase, phospholipase A, hyaluronidase, phosphodiesterase, RNase, DNase, leucine amino peptidase, amino acid esterase, and proteolytic enzymes were tested using purified toxin from *Enhydrina schistosa* venom, none of these activities were found. It was well studied that the neurotoxins from cobra and krait venoms blocked the neurotransmission at the postsynaptic site of neuromuscular junction (Lee, 1970). Sea snake venoms are also believed to inhibit the neuroactivity at the neuromuscular junction (Cheymol et al., 1967). All sea snake venom toxins have isoelectric points of above 9 (Toom et al., 1969; Tu and Toom, 1971; Tu et al., 1971; Tu and Hong, 1971). It is very likely that the basicity of the toxin causes the toxic action. Probably the negatively charged toxins firmly bind to the positively charged proteins in the postsynaptic site (acetylcholine receptor) or a neuromuscular junction. Thus the toxin produces a nondepolarizing neuromuscular block by action on the postjunctional membrane of the motor end plate without affecting the release of acetylcholine.

ACKNOWLEDGMENTS

I wish to express my sincere gratitude to my colleagues for their help. They are Dr. Paul M. Toom, Dr. Bor-Shyue Hong, J. D. Romer, Vice Admiral Ying Srihong, M. Lilabhan, and R. M. Durano. This investigation was supported by NIH Grant 5R01 GM 15591-04, GM 19172-03, and Career Development Award 5K04 GM 41768-03.

REFERENCES

Angeletti, R.H., Bichim. Biophys. Acta 214, 478 (1970).

Arai, H., Tamiya, N., Toshika, S., Shinonaga, S., and Kano, R., J. Biochem. (Tokyo) 56, 568 (1964).

Aloof-Hirsch, S., DeVries, A., and Berger, A., Biochim. Biophys. Acta 154, 53 (1968).

Botes, D.P., and Strydom, D.J., J. Biol. Chem. 244, 4147 (1969).

Botes, D.P., Toxicon 8, 125 (1970).

Chang, C.C., and Hayashi, K., Biochem. Biophys. Res. Commun. 37, 84 (1969).

Chao, L., and Einstein, E.R., J. Neurochem. 17, 1121 (1970).

Cheymol., J., Barme, M., Bourillet, F., and Roch-Arveiller, M., Toxicon 5, 111 (1967).

Endo, Y., Sato, S., Ishii, S., and Tamiya, N., Biochem. J. 122, 463 (1971).

Haile, N.S., Sarawak Mus. J. 8, 743 (1958).

Homma, M., Okonogi, T., and Shogi, M., Gumma J. Med. Sci. 13, 283 (1964).

Hong, B., "Studies on the Toxic Principles in the Venoms of Sea Snakes," Ph.D. Thesis, Colorado State University, 1971.

Hong, B., and Tu, A.T., Fed. Proc. 29,888 (1970).

Karlsson, E., Arnberg, H., and Eaker, D.L., Eur. J. Biochem. 21, 1 (1971).

Karlsson, E., Eaker, D.L., and Porath, J., Biochim. Biophys. Acta 127, 505 (1966).

Larsen, P.R., and Wolff, J., J. Biol. Chem. 243, 1283 (1968).

Lee, C.Y., Clin. Toxicol. 3, 457 (1970).

Lo, T.B., and Chang., Q., J. Chinese Chem. Soc. 13, 203 (1966).

Lo, T.B., and Chen, Y.H., J. Chim. Chem. Soc. 13, 195 (1966).

Mebs, D., Narita, K., Iwanaga, S., Samejima, Y., and Lee, C.Y.., Biochem. Biophys. Res. Comm. 44, 711 (1971).

Miranda, F., Kupeyan, C., Rochat, H., Rochat, C., and Lissitzky, S., Eur. J. Biochem, 17, 477 (1970).

Nakai, K., Nakai, C., Sasaki, T., Kakiuchi, K., and Hayashi, K., Naturwissenschaften 57, 387 (1970).

Narita, K., and Lee, C.Y., Biochem. Biophys. Res. Comm. 41, 339 (1970).

Porath, J., Mem. Inst. Butantan. Symp. Internat. 38, 379 (1966).

Romer, J.D., Mem. Hong Kong Natural History Soc. No.5 (1961).

Romer, J.D., "Illustrated Guide to the Venomous Snakes of Hong Kong with Recommendation for First Aid Treatment of Bites," Government Printer, Hong Kong, 1965.

Seto, A., Sato, S., and Tamiya, N., Biochim Biophys. Acta 214, 483 (1970).

Tamiya, N., and Arai, H., Biochem. J. 99, 624 (1966).

Taylor, E.H., "The Snakes of the Philippine Island," Bureau of Printing, Manila, 1922.

Taylor, E.H., University of Kansas Science Bulletin 45, No.9 (1965).

Toom, P.M., Squire, P.G.., and Tu, A.T., Biochim. Biophys. Acta 181, p. 339 (1969).

Tu, A.T., and Ganthavorn, S., Amer. J. Trop. Med. Hyg. 18, 151 (1969).

Tu, A.T., and Hong, B., J. Biol. Chem. 246, 2772 (1971).

Tu, A.T., and Toom, P.M., J. Biol. Chem. 246, 1012 (1971).

Tu, A.T., and Tu, T., in "Poisonous and Venomous Marine Animals of the World" Vol. 3, p.885. U.S. Government Printing Office, Washington, D.C., 1970.

Tu, A.T., Hong, B., and Solie, T.N., Biochemistry 10, 1295 (1971).

Tu, A.T., Passey, R.B., and Toom, P.M., Arch. Biochem. Biophys. 140, 96 (1970).

Uwatoko, Y., Nomura, Y., Kojima, K., and Obo, F., Acta Med. Univ. Kogoshima 8, 141 (1966a).

Uwatoko, Y., Nomura, Y., Kojima, K., and Obo, F., Acta Med. Univ. Kagoshima 8, 151 (1966b).

Uwatoko-Setoguchi, Y., Acta Med. Univ. Kagoshima 12, 73 (1970).

Yang, C.C., J. Biol. Chem. 240, 1616 (1965).

Note added in proof: Since this article was written in 1971, the author collected sea snakes in 1972 in Asia and in Central America in 1973. Research data resulting from these trips are listed below:

Raymond, M.L., and Tu, A.T., Biochim. Biophys. Acta 285, 498 (1972).

Tu, A.T., Ann. Rev. Biochem. 42, 235 (1973).

Tu, A.T., and Stringer, J.M., J. Herpetology 7, 384 (1973).

Tu, A.T., J. Agr. Food Chem 22, 36 (1974).

Tu, A.T., and Salafranca, E.S., Am. J. Trop. Med. Hyg. 23, 135 (1974).

Tu, A.T., Sea Snake Investigation in the Gulf of Thailand, J. Herpetology, in press (1974).

Chapter 13

MARINE PROSTAGLANDINS

Alfred J. Weinheimer

Department of Chemistry
University of Oklahoma
Norman, Oklahoma

The marine natural product program at Oklahoma has devoted much of its effort to studies of the readily accessible group of Caribbean invertebrates known as gorgonians (octocorals). These bushy and rather large colonial organisms are widely distributed throughout the Caribbean region, often occurring in dense growths on coral formations or other solid bottoms, where they are usually the most prominent members of the faunal community. Structurally, the gorgonian is comprised of a tough central proteinaceous stalk, which supports an outer cortical layer of softer material in which individual polyps are embedded. Distributed within certain portions of the polyp tissues are single-celled algae (zooxanthellae), which are photosynthetically active and which appear to play a significant role in the biochemical processes (Rice et al., 1970) of the gorgonian.

This group of organisms has been found to elaborate a wide variety of unusual natural products. These include a number of sesquiterpene hydrocarbons antipodal to those found in terrestrial sources (Weinheimer et al., 1967, 1968a); a series of closely related diterpene lactones, cembranolides, based on the 14-membered carbocyclic cembrane skeleton (Weinheimer et al., 1967, 1968b); a new C-30 sterol, gorgosterol (Hale et al., 1970; Ling et al., 1970), and several of its derivatives (Schmitz and Pattabhiraman, 1970; Enwall et al., 1972), which possess a unique cyclopropane containing side chain; and a series of compounds based on the bis lactone ancepsenolide

(Schmitz and Lorance, 1971; Schmitz et al., 1966, 1969). The occurrence of occasionally large quantities of terpenoid compounds in these organisms has prompted the speculation (Ciereszko et al., 1960) that they may be the products of algal metabolism within the colony.

Undoubtedly the most surprising and potentially the most significant discovery made in this program has been the recognition of large quantities of prostaglandins in the gorgonian *Plexaura homomalla* (Weinheimer and Spraggins, 1969). The prostaglandins are a group of new hormonelike materials, which display an amazing array of physiological effects in man and other mammals. With the exception of this gorgonian, they have not been observed in lower forms of life. A great deal of excitement exists over the promise that prostaglandins hold for many new types of therapy, and the realization of some of these hopes may begin in either 1972 or 1973 when the first product is expected to be marketed.

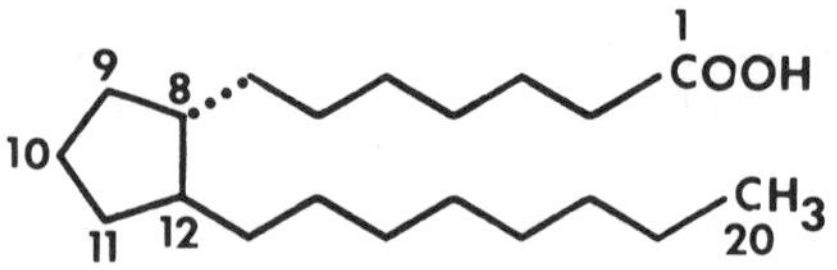

FIG. 1. Prostanoic acid.

Before proceeding further, it will be helpful to briefly review the structural relationships between the individual members of the group of prostaglandin molecules, and to define the "code" employed for simple reference to each of them. All the prostaglandins (PG's) are closely related structurally and may be considered to be derived from the parent 20-carbon monocyclic fatty acid, prostanoic acid (Fig. 1). The nature and arrangement of functional groups in the cyclopentane ring is designated by use of one of the letters: E, F, A, or B. An E type of prostaglandin (PGE) has a ketone function at C-9, together with an hydroxyl group at C-11. In natural PGE's, this hydroxyl invariably is disposed below the plane of the 5-membered ring as indicated by the dashed line (Fig. 2). Prostaglandins of the F type are characterized by two hydroxyl groups, one at C-9 and the other at C-11. PGF's may be regarded as the products of reduction of the 9-ketone function of PGE's. The disposition of the 9-hydroxyl below the plane of the ring is shown by a dashed line, and this is designated in the letter code with the subscript α, e.g., PGF_{α} (Fig. 2). The A type of prostaglandin has a 9-ketone function and a double bond between C-10

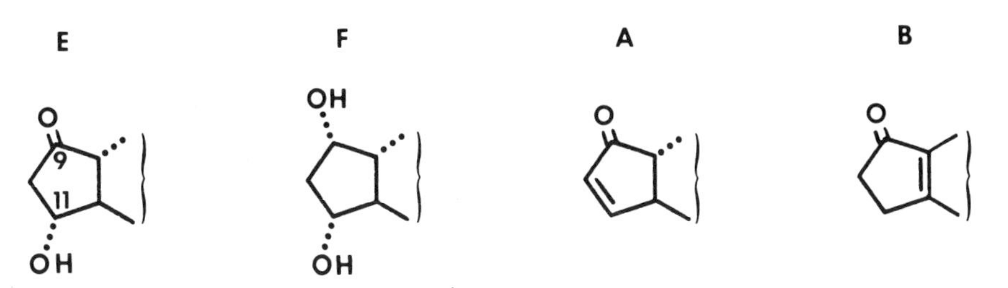

FIG.2. Distinguishing structural features of the cyclopentane rings of prostaglandins.

and C-11, and may be regarded as the dehydration product of a PGE. In fact, this dehydration is readily effected by treatment of a PGE with acid. Finally, the B type of prostaglandin possesses a 9-ketone function and a double bond between C-8 and C-12. A PGB results from treatment of a PGA with base and is generally, biologically inactive (Fig.2).

Attached to these ring systems are two aliphatic side chains. Disposed below the plane of the ring at C-8 is a 7-carbon side chain bearing a carboxylic acid function, and trans to this, i.e., disposed above the plane, is an 8-carbon side chain at C-12. The latter chain invariably possesses a trans double bond between C-13 and C-14, and an hydroxyl group at C-15, which is disposed downward with respect to the plane of carbon atoms as drawn in Fig.3. This (S) configuration at C-15 is common to all mammalian prostaglandins.

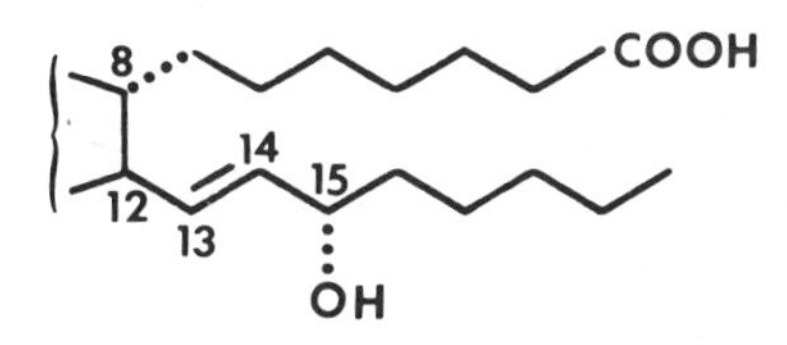

FIG.3. Prostaglandin side chains.

The code designation for PG's is completed with a numerical subscript indicating the number of double bonds occurring in the two side chains. For example, if only the 13,14 double bond is present in the side chains of a prostaglandin having an E ring pattern, the compound is designated as PGE_1 (Fig.4). The subscript 2 is employed to designate the presence of a second side-chain double bond, which occurs between C-5 and C-6 in the upper side chain and which has the cis configuration, as in PGA_2 and $PGF_{2\alpha}$ (Fig.4). A minor group of prostaglandins possesses a third double bond (cis) between C-17 and C-18.

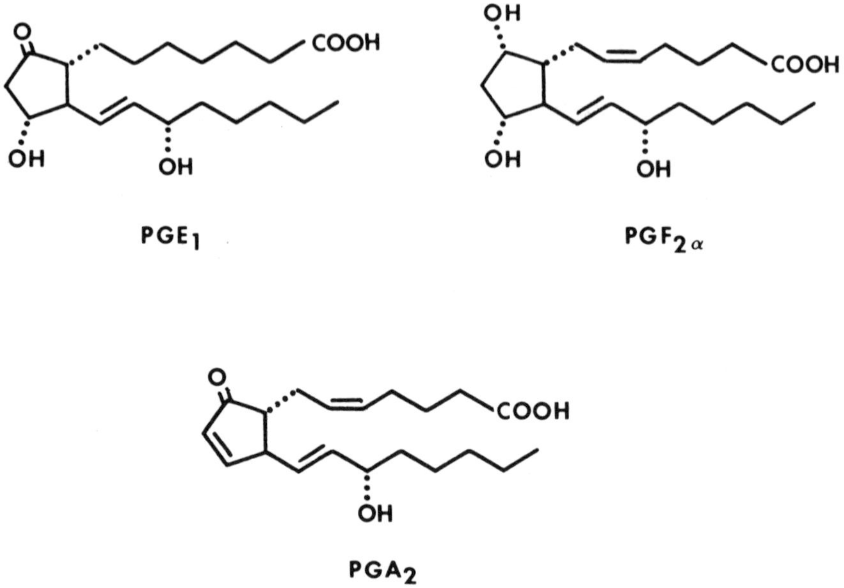

PGE_1

$PGF_{2\alpha}$

PGA_2

FIG.4. Typical prostaglandins.

Together, the various prostaglandins display a wide variety of intense biological activities, although a given effect is often most pronounced in one or another individual member of the group. A few examples (Hinman, 1970), selected from those which appear likely to be among the first practical applications, will illustrate the great potential they hold for medical use. Intravenous injection of PGE_1 inhibits blood platelet aggregation, offering promise for the prevention and treatment of thrombi and emboli. Similar application of PGA_1 inhibits gastric secretion and may find use in treatment of ulcers. Both PGA_1 and PGA_2 cause reduction in blood pressure and may be useful in treatment of certain types of hypertension. PGE_2 and $PGF_{2\alpha}$ are capable of inducing abortion in man up to the fifth month of pregnancy and are under intensive study as abortifacients. Finally, when inhaled in aerosol form, PGE_1 displays bronchodilation effects and shows great promise for prolonged relief of asthma symptoms.

The problem of inadequate supplies has plagued the study of the biological properties of prostaglandins from the outset. Microgram quantities of several prostaglandins are present in human seminal plasma, the source from which a number of the individual pure compounds were first isolated and characterized

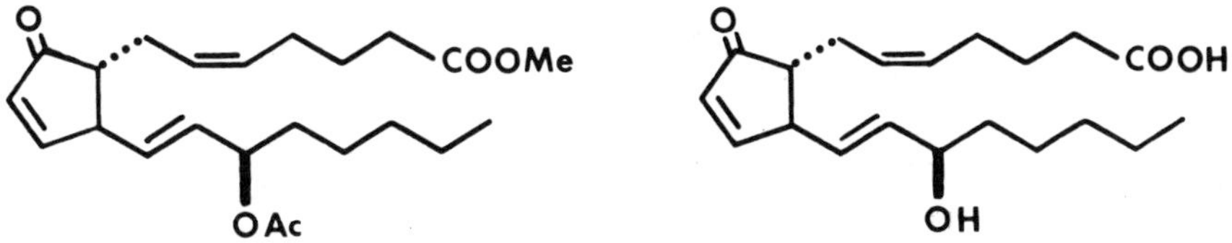

FIG. 5. 15-epi-Prostaglandins of Plexaura homomalla.

by Bergström and his collaborators in the early 1960's (Bergström, 1967). Gram quantities were subsequently prepared by enzymatic transformation of polyunsaturated acids using homogenates or acetone powders of ovine and bovine seminal vesicles (Hinman, 1967). More recently, larger quantities of prostaglandins have become available through practical methods of total synthesis. The sensitivity of prostaglandins to acid and base imposes severe constraints on synthetic approaches, and despite an apparent structural simplicity, they have posed a great challenge to the ingenuity of the synthetic chemist. The syntheses developed by E. J. Corey of Harvard have elegantly mastered these difficulties, and since 1970 all members of the series have been available by total synthesis (Corey, 1971).

In 1969, we reported (Weinheimer and Spraggins, 1969) our observation that large quantities of prostaglandin compounds were present in the gorgonian Plexaura homomalla from the Florida Keys. Although traces of related compounds were evident, the hexane extract of this organism consisted predominantly of two prostaglandins, which amounted to about 1.5% of the dry weight of the gorgonian. The major component (1.3%) was the methyl ester, acetate of 15-epi-PGA_2, and the minor component (0.2%) was the corresponding hydroxy acid, 15-epi-PGA_2, (Fig. 5). These coral prostaglandins differed from the mammalian form only in the configuration at C-15, the coral compounds having the (15 R) configuration as opposed to the epimeric (15 S) configuration of mammalian prostaglandins. The presence of so characteristically mammalian compounds in so distant a phylum raises many intriguing biological and evolutionary questions. Of all the gorgonians we have studied, Plexaura homomalla alone contains detectable quantities of prostaglandins. The closely related Plexaura flexuosa contains none, and the only other member of the genus, Plexaura nina, is extremely rare, only one specimen having ever been recorded (Bayer, 1961a). The function of these compounds in the gorgonian is still unknown.

This improbable source of prostaglandins was discovered before suitable total syntheses were fully developed, and the practical aspects of using the coral materials as a source of starting material for chemical conversion to normal prostaglandins were evaluated by a number of pharmaceutical firms. The details of one of these studies, by the Upjohn group, were disclosed at the prostaglandin conference in September, 1970 (Bundy et al., 1971). A series of three chemical steps was employed to effect epimerization at C-15 to give the normal PGA_2 derivative, which was transformed in several further steps to the more useful PGE_2 and $PGF_{2\alpha}$. The overall yield of $PGF_{2\alpha}$ from the major coral prostaglandin was 15%. This work demonstrated the practicability of using the coral compounds as starting material for synthesis of mammalian prostaglandins, and obviously opened for serious consideration the possible harvest of Plexaura homomalla as an alternative to total synthesis for production purposes.

In addition to economic considerations, which do not appear to be unfavorable, other factors will be important in determining the extent to which this approach can actually be practiced. To be inexpensive, collection on a large scale will require dense and homogeneous stands of the gorgonian. The stands will have to be extensive enough to support continuing harvest on the scale needed to meet annual production needs, and harvest practices will have to be compatible with the regrowth rate of the gorgonian.

Although no detailed census information is available for the Caribbean gorgonians, Plexaura homomalla is in fact widely distributed throughout the region (Bayer, 1961b), and frequently occurs in thick stands. Nothing is yet known of the growth rate of P. homomalla specifically, but an order of magnitude is suggested by the observation (Bayer, 1961c) that natural forces annually destroy one-fifth of the gorgonian fauna at one location in the Florida Keys.

Thus it appears to be well within the limits of feasibility that manufacture of prostaglandins for medicinal use could be based on the programmed harvest of this gorgonian. Whether or not this develops into a literal instance of "drugs from the sea," the discovery of prostaglandins in the marine environment has awakened interest in this vast and virtually untapped resource. It is already certain that the coming decade shall witness an intensified search for other physiologically active compounds from marine organisms.

Shortly after this Symposium was held, the Upjohn group reported (Schneider et al., 1972) a striking sequel to this account of marine prostaglandins. Apparently as a result of an extensive survey of many reef areas in the Caribbean, a strain of P. homomalla has been found, which, unlike the strain from the Florida Keys, contains normal PGA_2 derivatives and not 15-epi PGA_2 derivatives. Specifically, this strain contains 1.4% of the methyl ester of normal PGA_2, 0.4% of normal PGA_2, and trace amounts of normal PGE_2. These materials are readily convertible to PGE_2 and $PGF_{2\alpha}$ in good yields with minimum effort, and the new strain will obviously be the preferred source of prostaglandin compounds.

ACKNOWLEDGMENTS

I wish to acknowledge the able collaboration of Dr. Robert L. Spraggins in our work with the gorgonian prostaglandins. Were it not for his capability and persistence with these very sensitive compounds, this chapter of prostaglandin chemistry might not yet have been written. I also wish to acknowledge my colleagues, Professors Leon S. Ciereszko, Francis J. Schmitz, and Pushkar N. Kaul, whose collaboration has been so important in developing the marine natural products program at the University of Oklahoma. This work was supported by a grant from the National Heart Institute.

REFERENCES

Bayer, F.M., "The Shallow-Water Octocorallia of the West Indian Region," p.100. Martinus Nijhoff, The Hague, 1961a.

Bayer, F.M., "The Shallow-Water Octocorallia of the West Indian Region," p.99. Martinus Nijhoff, The Hague, 1961b.

Bayer, F.M., "The Shallow-Water Octocorallia of the West Indian Region," p.320. Martinus Nijhoff, The Hague, 1961c.

Bergstrom, S., Science 157, 382 (1967).

Bundy, G.L., Lincoln, F.H., Nelson, N.A., Pike, J.E., and Schneider, W.P., Ann. N.Y. Acad. Sci. 180, 76 (1971).

Ciereszko, L.S., Sifford, D.H., and Weinheimer, A.J., Ann. N.Y. Acad. Sci. 90, 917 (1960).

Corey, E.J., Ann. N.Y. Acad. Sci. 180, 24 (1971).

Enwall, E.L., van der Helm, D., Hsu, I.N., Pattabhiraman, T., Schmitz, F.J., Spraggins, R.L., and Weinheimer, A.J., Chem. Commun. 215 (1972).

Hale, R.L., Leclerq, J., Tursch, B., Djerassi, C., Gross, R.A., Weinheimer, A.J., Gupta, K.C., and Scheuer, P.J., J. Amer. Chem. Soc. 92, 2179 (1970).

Hinman, J.W., BioScience 17, 779 (1967).
Hinman, J.W., Prostagrad. Med. J. 46, 562 (1970).
Ling, N.C., Hale, R.L., and Djerassi, C., J. Amer. Chem. Soc. 92, 5281 (1970).
Rice, J.R., Papastephanou, C., and Anderson, D.G., Biol. Bull. 138, 334 (1970).
Schmitz, F.J., and Lorance, E.D., J. Org. Chem. 36, 719 (1971).
Schmitz, F.J., and Pattabhiraman, T., Chem. Commun. 6073 (1970).
Schmitz, F.J., Kraus, K.W., Ciereszko, L.S., Sifford, D.H., and Weinheimer, A.J., Tetrahedron Letters 97 (1966).
Schmitz, F.J., Lorance, E.D., and Ciereszko, L.S., Proceedings of Food-Drugs from the Sea Symposium, Marine Technology Society 315 (1969).
Schneider, W.P., Hamilton, R.D., and Ruhland, L.E., J. Amer. Chem. Soc. 94, 2122 (1972).
Weinheimer, A.J., Washecheck, P.H., van der Helm, D., and Hossain, M.B., Chem. Commun. 1070 (1968a).
Weinheimer, A.J., Schmitz, F.J., and Ciereszko, L.S., Transactions of the Drugs from the Sea Symposium, Marine Technology Society 135 (1967).
Weinheimer, A.J., Middlebrook, R.E., Bledsoe, J.O. Jr., Marsico, W.E., and Karns, T.K.B., Chem. Commun. 384 (1968b).
Weinheimer, A.J., and Spraggins, R.L., Tetrahedron Letters 5185 (1969).

AUTHOR INDEX

Underlined numbers give the page on which the complete reference is listed.

Abbott, B.C., 153, 154, 161, 171, 172
Abbott, R.T., 141, 148
Abram, R.G., 10, 12
Ackerman, D., 126, 134
Addicott, F.T., 201, 205
Adelman, W.J., 117, 121
Aitken, J. B., 198, 205
Aleem, A.A., 27, 34
Alender, C.B., 19, 20, 27, 32, 34, 96, 96, 97
Almodovan, L.R., 143, 148
Aloof-Hirsch, S., 226, 229
Altman, P.L., 143, 148
Amatniek, E., 10, 13
Anderson, D.G., 231, 238
Angeletti, R.H., 228, 229
Anthony, A., 199, 205
Arai, H., 213, 214, 224, 225, 230
Arnberg, H., 226, 229
Axelsson, J., 10, 12

Bagnis, R., 15, 16, 17, 18, 21, 22, 23, 26, 27, 28, 30, 31, 34, 35, 36
Ballantine, D., 153, 171
Banner, A.H., 15, 16, 17, 18, 19, 20, 21, 25, 26, 27, 30, 31, 32, 34, 35, 36
Bardach, J.E., 143, 148
Barme, M., 228, 229
Barnes, J.H., 1, 12, 127, 131, 135
Bartsch, A.F., 30, 34
Baslow, M.H., 15, 16, 17, 34, 36, 175, 182
Battey, Y., 16, 36
Baxter, E.H., 130, 131, 135
Bayer, F.M., 235, 236, 237
Bennett, I., 123, 137
Bennett, J., 28, 31, 35
Bentley, J.A., 199, 205
Beraldo, W.T., 37, 97
Berek, U., 52, 97
Berger, A., 226, 229
Bergmann, F., 154, 155, 159, 161, 171, 173
Bergmann, W., 193, 194
Bergstrom, S., 235, 237
Bernfeld, L., 148,, 148
Binford, J.S., Jr., 166, 169, 170, 171
Bledsoe, J.O., Jr., 231, 238
Blunden, G., 199, 205
Boisseau, J.P., 123, 135
Bolker, H.I., 193, 194
Bordes, F.P., 24, 36
Bornman, C.H., 201, 205
Botes, D.P., 226, 229
Bouder, H., 24, 34
Bouder, M.J., 24, 34
Bourillet, F., 228, 229
Brady, A.J., 11, 12
Brock, V.E., 26, 31, 34
Brody, S., 57, 96
Brown, P.A., 164, 170, 172
Buchwald, H.D., 10, 12, 115, 116, 118, 121, 148, 149
Bundy, G.L., 236, 237

Bunker, N.C., 19, 26, 35
Burg, E.A., 201, 205
Burg, S.P., 201, 205
Burke, J.M., 153, 171
Burkhalter, A., 49, 97
Burkholder, L.M., 143, 148
Burkholder, P.R., 143, 148
Burklew, M.A., 18, 35, 144, 148, 152, 153, 173
Burnett, J.W., 123, 124, 127, 129, 135, 137

Campbell, J.C., 15, 35
Campbell, J.E., 152, 172
Campbell, J.M., 83, 96
Cantacuzene, J., 126, 135
Cargo, D.G., 124, 127, 135
Carvalho, A.P., 11, 12
Castellani, B.A., 79, 97
Cavallo, A., 24, 34, 36
Cecil, J.T., 191, 194
Cha, Y.N., 119, 121
Chang, C.C., 226, 229
Chang, Q., 226, 229
Chanley, J.D., 183, 184, 194, 194
Chao, L., 227, 229
Chapman, G.B., 99, 113, 123, 135
Charney, J., 79, 96
Chatterjee, A.B., 153, 172
Chemier, G., 24, 34
Chen, Y.H., 229
Chesher, R.H., 31, 34
Cheymol, J., 228, 229
Childers, J.T., 198, 205
Chu, G.W., 27, 35
Ciereszko, L.S., 231, 232, 237, 237, 238
Cleland, J.B., 1, 12, 125, 131, 135
Cohn, G.H., Jr., 49, 97
Collier, A., 151, 153, 171
Connell, C.H., 153, 172
Cooper, M.J., 18, 19, 27, 30, 31, 32, 34
Corey, E.J., 235, 237
Coursen, B.W., 129, 132, 136
Crawford, J.E., 198, 206
Crone, H.D., 126, 130, 135, 136
Cross, J.B., 153, 172
Cuervo, L.A., 117, 221
Cummins, J.M., 152, 153, 172

Dafni, Z., 157, 158, 170, 172
Damato, A.N., 185, 195
Damboviceanu, A., 126, 135
Darwin, C., 126, 135
Davenport, D., 124, 135
Dawson, E.Y., 27, 34
Deguchi, T., 117, 121
Derting, C.W., 198, 206
Dessent, T.A., 164, 170, 172
DeSylva, D.P., 25, 34
Dettbarn, W.D., 118, 121
deVillez, E.J., 139, 140, 141, 146, 147, 149
DeVries, A., 226, 229
Dilaimy, M.S., 127, 129, 135
Dimitrov, G.D., 5, 12
Dittmer, D.S., 143, 148
Dixon, W.J., 176, 182
Djerassi, C., 231, 237, 238
Dodge, E., 99, 100, 103, 113, 125, 127, 129, 130, 132, 136
Duchemin, C., 1, 2, 3, 10, 12, 127, 130, 135
Dudeck, L.E., 175, 182
Durant, R.C., 183, 184, 194
Durham, L., 118, 121

Eaker, D.L., 226, 229
Eigner, E.A., 168, 172
Einstein, E.R., 227, 229
Eisenberg, R.S., 10, 12
Endean, R., 1, 2, 3, 9, 10, 11, 12, 127, 130, 131, 135
Endo, Y., 225, 229
Enwall, E.L., 231, 237

Fash, F.J., 183, 184, 194
Fawcett, D.W., 123, 137
Feigen, G.A., 46, 57, 83, 96, 96, 97
Fell, H.B., 192, 193, 194
Fellmeth, E.L., 16, 17, 36
Fischer, H.G., 10, 12, 115, 116, 118, 121, 148, 149
Florkin, M., 148, 148
Forbes, J.J., 16, 36
Frank, G.B., 10, 12
Fraser, E.H., 1, 2, 3, 10, 12 127, 130, 135
Fraser, I.M., 139, 143, 148
Freeman, S.E., 130, 135
Freygang, W.H., 10, 13
Frieden, C., 67, 97
Friess, S.L., 183, 184, 194
Fuchs, F., 5, 11, 12
Fuhrman, F.A., 10, 12, 115, 116, 117, 118, 119, 121, 148, 149

Gage, P.W., 10, 12
Ganthavorn, S., 221, 230
Gargus, J.L., 175, 182
Garriott, J.C., 134, 135
Gentile, J., 143, 149
Gergely, J., 5, 12
Giberman, E., 170, 172
Goeschl, H.D., 201, 206
Goldner, R., 127, 129, 135
Goldsmith, E.D., 184, 194
Gorham, P.R., 141, 148
Gross, R.A., 231, 237
Grundfest, H., 10, 13
Gunn, J.A., 3, 12
Gupta, K.C., 193, 194, 231, 237

Habekost, R.C., 139, 143, 148
Hadji, L., 96, 96
Haile, N.S., 210, 229
Hakluyt, R., 125, 135
Hale, R.L., 231, 237, 238
Halstead, B.W., 15, 16, 18, 19, 24, 25, 26, 27, 30, 34, 35, 139, 143, 147, 148, 152, 153, 172
Hamilton, R.D., 236, 238
Hand, C., 105, 113
Harada, R., 118, 121
Harris, J.B., 118, 119, 121
Hashimoto, Y., 15, 16, 17, 26, 27, 31, 35, 36, 192, 193, 195
Hastings, L.G., 132, 133, 135
Hayashi, F., 200, 205
Hayashi, K., 226, 229
Hayat, M. A., 101, 113
Helfrich, P., 15, 18, 19, 20, 21, 23, 24, 25, 26, 27, 30, 31, 32, 34, 35
Henderson, L., 1, 2, 10, 11, 12, 130, 131, 135
Herz, R., 11, 13
Hestrin, S., 155, 173
Heyl, M.G., 164, 170, 172
Higman, H.B., 118, 121
Hille, B., 117, 118, 121
Hines, K., 129, 132, 136
Hinman, J.W., 234, 235, 238
Hoffman, P.G., Jr., 170, 172
Holtz, F., 126, 134
Homma, M., 224, 229
Hong, B., 212, 213, 214, 216, 218, 225, 226, 228, 229, 230
Hong, S.K., 17, 36
Hoshino, S., 185, 195
Hossain, M.B., 231, 238
Hsu, I.N., 231, 237
Hudak, W.V., 184, 194
Huddart, H., 10, 11, 12
Hurley, R., 128, 136
Hutner, S.H., 25, 35
Hutton, R.F., 152, 172
Hyman, L.H., 99, 113, 128, 136, 193, 194

Ikemoto, N., 5, 12
Ingle, R.M., 152, 153, 172, 173
Ishii, S., 225, 229
Ivy, A.C., 21, 35
Iwanaga, S., 226, 229
Iwanzoff, N., 123, 136

Jackim, E., 143, 149
Jacobsen, S., 68, 97
Jakowska, S., 175, 182, 183, 184, 195
Jones, A.C., 152, 172
Jones, E.C., 123, 136
Jones, R.L., 200, 205
Jones, R.S., 26, 31, 34

Kakiuchi, K., 226, 229
Kamiya, H., 16, 17, 35
Kankonkar, R.C., 5, 12
Kano, R., 213, 229
Kao, C.Y., 10, 12, 115, 116, 117, 118, 119, 120, 121
Karlsson, E., 226, 229
Karns, T.K.B., 231, 238
Keen, T.E.B., 126, 130, 135, 136
Keenan, J.P., 21, 35
Kellaway, C.H., 37, 97
Kendall, F.E., 53, 97
Kerr, S.E., 170, 172
Keynes, R.D., 117, 121
Kidron, M., 159, 161, 171, 172
Kingston, C.W., 2, 12
Kleinhaus, A.L., 119, 121
Kline, E.L., 128, 136
Kline, E.S., 128, 136
Kojima, K., 213, 214, 225, 230
Konosu, S., 16, 17, 35
Kosaki, T.I., 16, 17, 36
Kraus, K.W., 238
Kupeyan, C., 214, 226, 229

Ladue, K.T., 183, 195
Lallier, R., 190, 194
Lalone, R.C., 139, 140, 141, 146, 147, 149
Lane, C.E., 99, 100, 103, 113, 125, 127, 128, 129, 130, 132, 133, 134, 135, 136, 137
Lane, W.R., 130, 131, 135
Langer, G.A., 11, 12
Larsen, J.B., 129, 132, 133, 134, 135, 136
Larsen, P.R., 226, 229
Larson, E., 139, 140, 141, 146, 147, 149
Lasley, B.J., 184, 185, 195
Layne, E.C., 124, 127, 135
Leclerq, J., 231, 237
Lee, C.Y., 226, 228, 229
Legato, M.J., 11, 12
Lemon, J., McW., 139, 149
Lenhoff, H.M., 128, 136
Lewis, K.H., 15, 35, 152, 172
Li, K.M., 17, 35
Lichter, W., 175, 182
Lilleheil, G., 132, 136, 137
Lincoln, F.H., 236, 237
Ling, N.C., 231, 238
Lipsius, M.R., 119, 121
Lissitzky, S., 214, 226, 229
Lo, T.B., 226, 229
Loftfield, R.B., 168, 172
Lorance, E.D., 232, 238
Lotka, A.J., 57, 97
Lucas, A.H., 175, 182
Lynch, J.M., 116, 121

McCalla, D.R., 199, 201, 206
McClintock, D.K., 79, 97
McColm, D., 1, 2, 3, 10, 12, 127, 130, 135
McFarren, E.F., 15, 30, 34, 35, 147, 149, 152, 172
McLaughlin, J.J.A., 25, 35, 153, 154, 171, 172
Malarde, L., 28, 31, 35
Marchisotto, J., 153, 171
Markus, G., 79, 83, 96, 96, 97
Marr, A.G.M., 130, 135

Marr, H.G., 130, 131, 135
Marsico, W.E., 231, 238
Martin, D.F., 153, 154, 160, 161, 162, 163, 164, 165, 166, 167, 169, 170, 171, 172
Martin, E.J., 127, 136
Martin, J.A., 198, 205, 206
Mason, L.E., 32, 35
Matsumoto, K., 116, 121
Mebs, D., 226, 229
Mendes, J., 71, 97
Mezzetti, T., 183, 184, 194
Middlebrook, R.E., 231, 238
Miles, P.S., 15, 23, 30, 35
Miller, C.O., 199, 201, 205
Miranda, F., 214, 226, 229
Moikeha, S.N., 27, 35
Moore, J.W., 115, 117, 121
Moore, R.E., 193, 195
Morelon, R., 24, 28, 35
Morice, J., 24, 35
Morton, R.A., 18, 35, 144, 148
Mosher, H.S., 10, 12, 115, 116, 118, 121, 148, 149
Mowat, J.A., 198, 205
Murashige, T., 200, 201, 206
Murray, P.M., 184, 195

Nachmansohn, D., 118, 121
Nagasawa, J., 119, 120, 121
Nakai, C., 226, 229
Nakai, K., 226, 229
Nakamura, A., 5, 12
Nanai, F., 28, 31, 35
Narahashi, T., 115, 117, 121
Narita, K., 226, 229
Nash, J.B., 153, 173
Nelson, N.A., 236, 237
Nelson, W.R., 152, 153, 173
Niaussat, P., 24, 28, 35
Nielsen, C., 57, 96
Nigrelli, R.F., 175, 182, 183, 184, 185, 187, 188, 189, 191, 192, 194, 195
Nishiyama, A., 117, 118, 119, 120, 121
Noble, M., 9, 12
Nolan, S.F., 184, 195
Nomura, Y., 213, 214, 225, 230

Oates, K., 11, 12
Obo, F., 213, 214, 225, 230
Oesterling, R., 116, 121
Ohkubo, Y., 117, 121
Okihiro, M.M., 21, 35
Okonogi, T., 224, 229
Osborne, D.J., 199, 201, 206
Osburg, H.E., 184, 194

Padilla, G.M., 153, 154, 160, 161, 162, 163, 164, 165, 166, 167, 168, 169, 170, 171, 172
Papastephanou, C., 231, 238
Parker, G.H., 124, 136,
Parnas, I., 154, 155, 171, 172
Passey, R.B., 216, 218, 221, 230
Paster, Z., 158, 161, 171, 172
Pattabhiraman, T., 231, 237, 238
Pennell, R.B., 53, 60, 97
Perret, A., 126, 137
Peyrin, A., 24, 36
Pfeffer, R.A., 96, 96
Phillips, J.H., Jr., 127, 128, 136
Phinney, B.O., 200, 206
Phisalix, M., 124, 136
Picken, L.E.R., 100, 103, 113, 123, 136
Pierce, L.H., 124, 127, 135
Pike, J.E., 236, 237
Piyakarnchana, T., 15, 23, 30, 34, 35
Plaister, T.H., 143, 149
Ponder, E., 166, 172
Porath, J., 226, 229
Portier, P., 126, 136, 137
Prado, J.L., 68, 71, 97

Prakash, A., 153, 172
Pratt, H.K., 201, 206
Provasoli, L., 153, 171

Quaglio, N.D., 184, 195

Randall, J.E., 15, 16, 17, 18, 25, 26, 27, 28, 29, 30, 31, 35, 36, 145, 149
Ranzi, S., 190, 195
Rapoport, H., 116, 121
Rappaport, L., 200, 205
Raskova, H., 3, 13
Rauckman, B., 154, 161, 168, 172
Ray, S.M., 153, 173
Rayner, M.D., 16, 17, 35, 36
Redfern, P., 118, 121
Reich, K., 154, 155, 159, 161, 171, 173
Reiner, J.M., 168, 173
Reinwein, H., 126, 134
Reisfeld, R.A., 177, 182
Ricciutti, M.A., 185, 195
Rice, J.R., 231, 238
Richet, C., 126, 136, 137
Rio, G.J., 184, 187, 192, 195
Ritchie, J.M., 117, 121
Robson, E.A., 125, 137
Rocha e Silva, M., 37, 97
Roch-Arveiller, M., 228, 229
Rochat, C., 214, 226, 229
Rochat, H., 214, 226, 229
Rojas, E., 117, 121
Romer, J.D., 210, 228, 229
Rosa, R.C., 71, 97
Rosenberg, P., 118, 121
Rosenberg, R.F., 154, 155, 157, 173
Rosenfeld, G., 37, 97
Ross, D.M., 124, 135, 137
Rossi, C., 184, 194
Rounsefell, G.A., 152, 153, 173
Ruggieri, G.D., 184, 187, 188, 189, 191, 192, 194, 195
Ruhland, L.E., 236, 238
Russell, F.E., 10, 12, 15, 36
Ryther, J.H., 151, 173

Samejima, Y., 226, 229
Sanz, E., 96, 96, 97
Sasaki, S., 18, 27, 32, 34
Sasaki, T., 226, 229
Sato, S., 225, 226, 230
Schantz, E.J., 116, 121, 153, 173
Scheuer, P.J., 15, 16, 17, 18, 27, 32, 34, 36, 193, 194, 195, 231, 237
Schmitz, F.J., 231, 232, 237, 237, 238
Schneider, W.P., 236, 237
Scott, W., 117, 121
Senn, T.L., 198, 205, 206
Setliff, J.A., 17, 36
Seto, A., 226, 230
Shanes, A.M., 10, 13
Shapiro, B.I., 126, 132, 136, 137
Shaw, S.W., 20, 34
Shaw, T.I., 117, 121
Shilo, M., 153, 154, 155, 157, 158, 171, 172, 173
Shimada, S., 192, 195
Shinonaga, S., 213, 229
Shogi, M., 224, 229
Shore, P.A., 49, 97
Siegman, M.J., 119, 121
Sifford, D.H., 232, 237, 238
Sigel, M.M., 175, 182
Silva, F.J., 15, 35, 152, 172
Siman, R., Jr., 141, 146, 147, 149
Singh, H., 193, 195
Sirois, J.C., 198, 200, 206
Skaer, R.J., 100, 103, 113, 123, 136
Skelton, B.J., 198, 206
Skoog, F., 200, 201, 206
Slautterback, D.B., 103, 113

Slautterback, D.L., 123, 137
Sobotka, H., 183, 184, 194
Solie, T.N., 216, 218, 225, 226, 228, 230
Southcott, R.V., 1, 2, 12, 125, 131, 135
Spiegelstein, M.Y., 119, 120, 121, 154, 172
Spikes, J.J., 153, 173
Spraggins, R.L., 231, 232, 235, 237, 237, 238
Spurr, A.R., 101, 114, 201, 205
Squire, P.G., 215, 228, 230
Sreter, F.A., 5, 12
Steidinger, K.A., 152, 153, 173
Stempien, M.F., Jr., 184, 187, 192, 194, 195
Stephenson, N.R., 153, 173
Stevens, A.C., 152, 172
Stillway, L.W., 129, 137
Stone, J.H., 124, 127, 135
Stone, J.S., 127, 129, 135
Stotz, E.H., 148, 148
Street, H.E., 199, 205
Strydom, D.J., 226, 229
Styles, T.J., 185, 195
Sullivan, T.D., 183, 195
Sutherland, G.B., 83, 96
Sutton, J.S., 123, 124, 127, 135, 137
Sutton, L., 124, 135, 137
Suzuki, T., 119, 121
Svihovec, J., 3, 13

Taft, C.H., 147, 149
Takahashi, W., 15, 17, 36
Takata, M., 117, 121
Tamiya, N., 213, 214, 224, 225, 226, 230
Tanabe, H., 15, 35, 152, 172
Tanaka, M., 192, 193, 195
Tapu, J., 28, 31, 35
Taylor, E.H., 207, 209, 212, 230
Taylor, F.J.R., 153, 172
Taylor, R.E., 10, 13
Thesleff, S., 10, 12, 118, 119, 121
Thimann, K.V., 200, 206
Thompson, T.E., 123, 137
Tomarelli, R.M., 79, 96
Tomita, J.T., 96, 96, 97
Tonge, J.I., 16, 36
Toom, P.M., 215, 216, 218, 219, 221, 225, 226, 228, 230
Toshika, S., 213, 229
Trapani, I.L., 46, 97
Trembley, A., 125, 137
Tressler, D.K., 139, 149
Trethewie, E.R., 37, 97
Trief, N.M., 153, 173
Tsutsumi, J., 15, 17, 36
Tu, A.T., 37, 97, 207, 212, 213, 214, 215, 216, 218, 219, 221, 225, 226, 228, 229, 230
Turlapaty, P., 175, 182
Turner, R.J., 130, 135
Tursch, B., 231, 238

Ulitzur, S., 154, 156, 158, 171, 173
Urakawa, N., 117, 121
Uwatoko-Setoguchi, Y., 213, 214, 225, 230

Vaillant, A., 24, 36
Van Alstyne, M.A., 124, 136
van der Helm, D., 231, 237, 238
van Heukelem, 31, 34
Varner, J.E., 200, 205
Vayada, G., 116, 121
Veltri, A.M., 184, 195

Washecheck, P.H., 231, 238
Waugh, W.H., 3, 13
Weber, A., 11, 13
Weems, H.B., 184, 194
Weill, R., 99, 114
Weinheimer, A.J., 231, 232, 235, 237, 238

Wellham, L.L., 175, 182
Welsh, J.H., 126, 137
Werkheiser, W.C., 79, 97
Werle, E., 52, 97
Westfall, J.A., 103, 114, 123, 124, 137
Weyl, H., 105, 114
Wiberg, G.S., 153, 173
Williams, D.E., 177, 182
Williams, R.H., 139, 140, 149
Wilson, W.B., 15, 35, 152, 153, 173
Wilson, W.O., 143, 149
Winegrad, S., 11, 13
Wolff, J., 226, 229
Wong, J.L., 116, 121
Woodcock, A.H., 152, 173
Woods, D.L., 199, 205

Yamanouchi, T., 183, 195
Yanagita, T.M., 124, 137
Yang, C.C., 226, 230
Yariv, J., 155, 173
Yashida, T., 16, 17, 35
Yasumoto, T., 15, 16, 17, 26, 27, 31, 35, 36, 192, 193, 195
Yoshida, T., 15, 17, 34, 36

Zahl, P.A., 183, 195

SUBJECT INDEX

Abscissic acid, 200, 201
Acanthaster, 31, 186–191
Acanthuridae, 23, 25, 28
Acanthurus lineatus, 17, 26
Acanthurus xanthopterus, 25
Acetylcholine, 47, 49, 92, 134, 154, 228
Acetylcholinesterase, 228
Acrochordus granulatus, 207–209
Actineoplastic activity, 177, 179, 181
Actinia equina, 126
Actinopyga agassizi, 183, 189
Africa, 226
Agkistrodon, rhodostoma, 215, 216
Aipysurus eydouxi, 207–210, 214
Algae, 26, 29, 139–144, 197, 231, 232; see also Blue-green algae
Amino acid esterase, 220, 228
L-Amino acid oxidase, 3, 5, 9 10
Amylase, 200
Anaphylaxis, 126
Anemonia sulcata, 126
Animalization, 189, 190
Anticholinesterase, 17, 175, 177–179, 181, 228
Arbacia punctulata, 187, 188
Arius felis, 140
Ascophyllum nodosum, 197
Astichopus multifidus, 186, 188–192
Astrotia stokesii, 209
Atropine, 46, 47, 49, 134
Aurelia aurita, 128
Auxins, 198–200
Avena coleoptile, 200

Balanus nigrescens, 131
Barnacle, 10, 11, 131
Barracuda, 25
Bikini Atoll, 24, 32
Blue-green algae, 29, 143, 197, 198
Brachidontes exustus, 141, 145, 146
Bradykinin, 37, 38, 45, 46, 51, 52, 54, 55, 60, 61, 63, 65, 67–71, 73–79, 82–84, 87, 89, 92
D-Bromolysergic acid, 46, 47, 49
Bryopsis hypnoides, 140–145
Bufo marinus, 3
Bungarus multicinctus, 226

Caffeine, 5, 9–11
Calliactis parasitica, 124
Candida albicans, 191
Carangidae, 23, 30
Carcharinus menisorra, 15
amblyrhynchos, 16
Cardisoma guanhumi, 133, 134
Caribbean Sea, 16, 231, 236, 237
Cat, 130
Catecholamines, 120
Chelation, 197
Chicks, 140, 141, 143–146
Chione cancellata, 141, 145, 146
Chironex fleckeri, 1–11, 125, 127, 130–132
Cholinesterases, 37
Christmas Island, 25
Chrysaora quinquecirrha, 129
Chymotrypsin, 83, 84, 86, 89, 94
Ciguatera, 15–33

Ciguatoxin, 15-33
Clams, 92, 152
Clypeaster rosaceus, 186-188
Cnidoblasts, 99, 103, 123
Cnidocils, 103, 123, 124
Cobra, 214, 226, 228
Cocaine, 117
Coelenterata, 99-113, 123-134
Condylactis, 103, 126, 127
Corynactis viridis, 100, 103, 123
Crab, 38, 90, 91, 93-95
Crayfish, 11
Crown-of-thorns, 31, 186-191
Ctenochaetus striatus, 16, 17, 26, 28, 31
Curare, 133
Cyanea capillata, 126
Cyanophyta, 25, 27
Cytokinins, 200, 201

Deoxyribonuclease, 219
Diadema antillarum, 186-188
Dinoflagellates, 116, 151-173
Dog, 3, 130, 132, 133

Echidna, 130
Echinoderms, 37-97, 183-195
EDTA, 8
Elapidae, 226
Emydocephalus ijimae, 209
Enhydrina schistosa, 207-210, 214-220, 223, 225
Eniwetok Atoll, 24, 30
Enteromorpha, 139
Epinephelus fuscoguttatus, 17
Ethylene, 200, 201
Excitable membranes, 115

Fiji Islands, 31
Fish kill, 151-154
Flatworms, 123, 124
Florida, 18, 235-237
Florida Keys, 235-237
Formosa, 207, 210, 214, 226
Frog, 4, 11, 133, 134
Fruit shelf life, 198
Fundulus heteroclitus, 187

Gambusia affinis, 155
Geranium leaves, 205
Gibberellin, 198, 200
Gilbert and Ellis Islands, 18, 19, 24, 29, 30, 32
Glaucus marinus, 123, 125
Globulins, 53, 56, 57, 59-61, 63-68
Goat, 130
Golgi material, 123
Gonyaulax acatenella, 153
G. catanella, 116, 152, 153
G. moniliata, 153
G. tamarensis, 152, 153
Gonyaulax toxin, 116
Gorgonians, 231-238
Gracilaria compressa, 140-145
Great Barrier Reef, 18
Groupers, 16
Growth regulators, 198-200
Guam, 31
Guanidinium, 116
Guinea pig, 3, 5, 6, 44, 46, 48-52, 54, 60, 62, 67, 77, 78, 83, 134, 156
Gymnodinium breve, 153, 168, 170
Gymnodinium venificum, 153
Gymnothorax javanicus, 15, 19, 20, 26

Halistemma rubrum, 123
Haliclona viridis, 175-182
Halitoxin, 175
Hamsters, 140-142, 144
Hatching enzyme, 190
Hawaii, 21, 25, 29, 32, 38, 39
Hemachatus hemachatus, 226
Hemolysis, 153-164, 166-171, 175, 179, 180, 221
Hermit crab, 124
Herring, 18
Histamine, 38, 45-47, 49-51, 131, 134
Holothuria arenicola, 186, 188
H. densipedes, 186, 188
H. mexicana, 186, 188, 191
H. parvula, 186, 188, 191
H. vagabunda, 183
Holothurin, 183-192

Homarine, 126
Hong Kong, 207
Horse, 130
Hyaluronidases, 37, 219, 228
Hydra, 103, 125
Hydrophiidae, 215, 225
Hydrophis cyanocinctus, 207-210, 214, 221, 223
H. brooki, 209
H. caerulescens, 209
H. fasciatus, 209
H. klossi, 209
H. mamillaris, 209
H. melanocephalus, 209
H. ornatus, 207-210, 214
H. spiralis, 207, 209, 210
H. torquatus, 209
Hyla caerulea, 4
Hypotensive agents, 119, 120

India, 226
Indian Ocean, 16
Indole-3-acetic acid, 200
Indonesia, 18
Insecticides, 24, 28

Jaluit Atoll, 20, 24, 33
Japan, 207, 210, 212-214, 216-219, 224, 225
Johnson Island, 16, 20, 24, 30

Kallidin, 89
Kallikrein, 52, 67, 68
Kerilia jerdonii, 207-210
Killifish, 187, 188
Kinin, 52, 55, 64, 67-69, 71-73, 76, 77, 80-83, 89
Kininases, 67, 68, 70
Kolpophis annandalei, 209
Krebs-2 ascites tumors, 183

Lactuca sativa, 140, 143-145
Lapemis hardwickii, 207-211, 214, 216-221, 223, 225
Laticauda affinis, 209
L. colubrina, 209
L. laticaudata, 209
L. semifasciata, 209, 214, 216-220, 224-226
Leucine aminopeptidase, 220,228
Leucocytes, 185
Ligustrum lucidum, 201
Line Islands, 16
Loblolly pine seed, 202,203
Lobster, 38
Lutjanus bohar, 15, 16, 20-23, 25, 30
Lyngbya majuscula, 27

Malayan pit viper 216
Malaysia, 207, 210, 214-218, 220, 221, 225
Man-of-war fish, 125
Marquesas Island, 16
Marshall Islands, 24, 30, 32, 33
Mellita quinquiesperforata, 186-188, 191
Meoma ventricosa, 186, 188
Metridium dianthus, 126
Metridium senile, 127
Microcephalophis gracilis, 207, 208, 210
Microflora, 27, 33
Millepora alcicornis, 126
Millepora complanata, 126
Milleporina, 126
Monkey, 130
Monotoxis grandoculis, 18
Moray eel, 16, 17, 19, 20, 24, 26
Mouse, 5, 8, 38, 44, 51, 52, 90, 91, 93-95, 130-132, 140-142, 146
Mururoa Island, 22
Mussels, 152, 153

Naja haje, 226, 228
N. nigricollis, 226
N. naja, 226
N. nivea, 226
Nandina domesticum, 202
Nemaster rubiginosa, 186-188, 190, 191
Nematocysts, 1-11, 99-113, 123-134
New Hebrides, 21, 32
Newts, 116

Nickel, 24, 25
Nomeus gronovii, 125
Nucleases, 37
Nudibranchiata, 123

Ophiocoma echinata, 186, 188, 191
Oubain, 133, 134

Pachygrapsus crassipes, 90
Pacific Ocean, 16
Pagurus, 124
Palmyra Island, 16, 33
Palolo worm, 25
Paralytic shellfish poisoning, 116, 120
Paramecium, 132
Parrot fishes, 16, 29
Pedicellariae, 38-96
Pelamis platurus, 207, 209, 210, 214, 223
Phenylbutazone 47
Philippines, 18, 207, 209, 210, 212-214, 225
Phoenix Islands, 29, 30, 32
Phosphodiesterases, 37, 219, 228
Phospholipase A, 3, 10, 37, 130, 131, 216-219, 221, 227, 228
Physalia physalis, 99-114
Plant hormones, 198-200
Plexaura homomalla, 232, 235-237
Poisonous fishes, 15-36, 139-149; see also Paralytic shellfish poisoning, Shellfish poisoning
Porpita, 128
Portuguese man-o-war, 99-114
Praescutata viperina, 208, 210
Procaine, 2, 4, 10, 117
Prostaglandins, 231-238
Prostanoic acid, 232
Proteolytic enzymes, 37, 127, 220, 228
Prymnesin, 153-171
Prymnesium parvum, 153
Puffer fishes, 18, 25, 139-149
Purkinje fibers, 185, 186
Pyribenzamine, 46, 47, 49

Rabbit, 3, 6, 9, 38, 88, 93, 130, 159, 160, 162, 169, 170
Rat, 1, 3, 4, 6, 7, 9,44, 52, 54, 60, 62, 65, 67, 70, 71, 77, 78, 90, 131-134, 161, 185
Red tide, 151-173
Rhodactis howesii, 127
Ribonuclease, 220, 228
Ryukyu, 16, 27

Saccharomyces cerevisiae, 187, 191
Saipan, 31
Salamandridae, 116
Samoa, 31
Sarawak, 210
Sarcoma, 180, 183
Sardines, 18
Saxidomus giganteus, 116
Saxitoxin, 25, 115-120, 143, 153
Scaridae, 23, 28, 30
Schizothrix calcicola, 27
Sea anemone, 102, 124, 125, 126, 127
Sea cucumbers, 183, 187, 189, 192
Sea snake venom, 207-230
  amino acids in, 217
  chemical modification, 218, 220
  chemistry of, 215
  enzymes in, 219, 220
  immunology of, 221-224
  isolation of, 216
  mode of action, 228
  properties of, 216, 217
  purification of, 216, 225, 226
  toxicology of, 214
Sea snakes, 207-230
Sea urchin embryos, 184, 187, 188, 189, 190

Sea urchin toxin, 37-97
Seaweed extracts, 139, 197-206
Seed germination, 198, 201-203
Seriola dumerili, 16
Serotonin, 45, 46, 92
Serranidae, 23
Shark, 18, 26
Sheep, 130, 170
Shellfish poisoning, 152, 153; see also Paralytic shellfish poisoning
Siphonophores, 99
Sipunculid, 38
Snake venom, 37, 38, 207-230; see also Sea snake venom
Snappers, 16
Society Islands, 16
Sodium channels, 115, 117-119, 132, 133
Soft corals, 231-238
Sphoeroides maculatus, 139
S. nephelus, 139-148
S. testudinus, 140-148
Sphyraena barracuda, 16
Sponge, see Haliclona viridis
Staphlococci, 184
Stichopus japonicus, 192
Stomphia coccinea, 124
Strait of Malacca, 215, 218, 225
Surgeonfishes, 16

Tahiti, 16, 18, 22, 27, 28, 31, 32
Tarawa, 32
Taricha torosa, 116
Tarichatoxin, 116
Tetrahymena, 132
Tetramine, 126
Tetraodontidae, 18, 116, 147
Tetrodotoxin, 2, 4, 9, 10, 17, 24, 25, 115-121, 147, 148
Thailand, 207-210, 214, 216-218, 220, 221, 226
Thalassophis anomalus, 208, 210
Thalassophis viperina, 209
Tivela stultorum, 92
Toad, 3, 10
Tobacco leaves, 203, 204
Toxins, see specific compounds
Tremoctopus violaceus, 123
Trigoneline, 126
Tripneustes gratilla, 38
Trout, 130
Trypanosomes, 185
Tuamotu Islands, 18, 21, 22, 23

Ulva lactuca, 140-145
Urocanylcholine, 126

Vasoconstriction, 120
Vasodilation, 119
Vegetalization, 190
Velella, 128

Washington Island, 30

Zooxanthellae, 25, 231